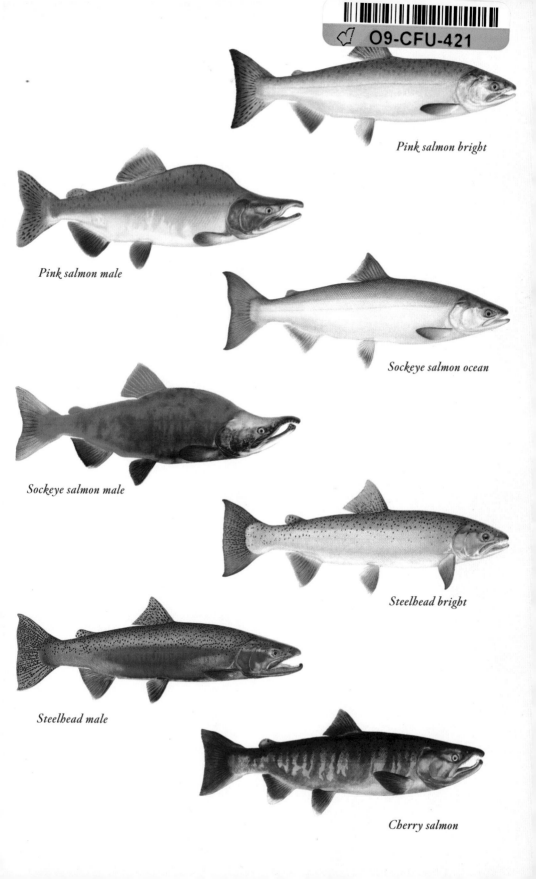

Pink salmon bright

Pink salmon male

Sockeye salmon ocean

Sockeye salmon male

Steelhead bright

Steelhead male

Cherry salmon

BY TUCKER MALARKEY

Stronghold

An Obvious Enchantment

Resurrection

STRONGHOLD

STRONGHOLD

One Man's Quest to Save the
World's Wild Salmon

TUCKER
MALARKEY

SPIEGEL & GRAU

NEW YORK

Published in the United States by Spiegel & Grau, an imprint of
Random House, a division of Penguin Random House LLC, New York.

SPIEGEL & GRAU and colophon is a registered
trademark of Penguin Random House LLC.

LIBRARY OF CONGRESS CATALOGING-IN-PUBLICATION DATA
Names: Malarkey, Tucker, author.
Title: Stronghold: one man's quest to save the world's wild salmon /
by Tucker Malarkey.
Description: New York: Spiegel & Grau, [2019]
Identifiers: LCCN 2018047731l ISBN 9781984801692 (hardback)
l ISBN 9781984801708 (ebook)
Subjects: LCSH: Rahr, Guido. l Salmon fishing. l Salmon—Conservation.
l Environmentalists—United States—Biography.
Classification: LCC SH684 .M345 2019 l DDC 639.2/756—dc23
LC record available at https://lccn.loc.gov/2018047731

Printed in the United States of America on acid-free paper

randomhousebooks.com
spiegelandgrau.com

2 4 6 8 9 7 5 3 1

First Edition

ENDPAPERS: Cherry salmon illustration by Kate Spencer;
all other illustrations by Joseph Tomelleri
BLACK-AND-WHITE ILLUSTRATIONS: G. Rahr

Book design by Caroline Cunningham

For Guido, for sharing your stories and your life,
and for our children—may they too fight for
what they love in this world

Wilderness is a necessity. . . . They will see what I meant in time.

—John Muir

CONTENTS

stronghold: a place that has been fortified in order to protect it against attack; a place where a cause or belief is defended or upheld; a bastion, a haven.

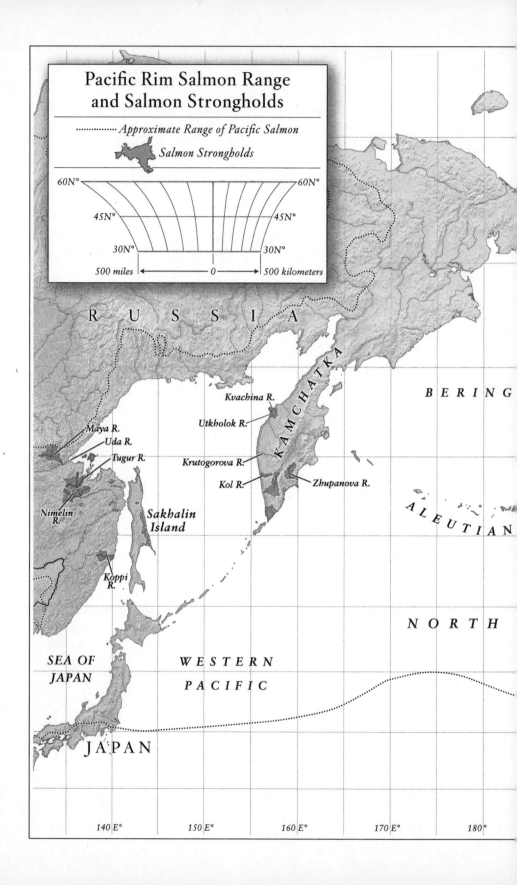

Pacific Rim Salmon Range and Salmon Strongholds

·············· *Approximate Range of Pacific Salmon*

Salmon Strongholds

60N°   60N°

45N°   45N°

30N°   30N°

500 miles |← →| 500 kilometers
0

RUSSIA

KAMCHATKA

BERING

Kvachina R.

Utkholok R.

Maya R.
Uda R.
Tugur R.

Krutogorova R.

Kol R.   Zhupanova R.

Nimelin R.

Sakhalin Island

ALEUTIAN

Koppi R.

NORTH

SEA OF JAPAN

WESTERN

PACIFIC

JAPAN

140 E°   150 E°   160 E°   170 E°   180°

INTRODUCTION

Kamchatka, Russia, 1999

T HE MI-8 WAS PACKED beyond all reasonable capacity. Thirty-two passengers, two zodiacs with outboard jet engines, three whitewater rafts, six wall tents, scientific equipment, rods, food, and miscellaneous gear filled the chopper from floor to ceiling. The only way back to civilization was over the Sredinny mountain range, whose peaks rose to more than eleven thousand feet. The pilot tapped his watch to urge one last passenger on board, but Guido Rahr refused. Inside the helicopter the group was getting restless. Their fuel was limited and the weather was shifting. What was the holdup? The men conferred loudly over the thrum of the whirling rotor. What was going on?

In bush flying, it's said that if you can get the doors of a helicopter closed, it will fly. The Mi-8 had two powerful turbine engines and was built for the toughest conditions. But with the weight of the gear and a section of heavy oil-drilling pipe, the chopper's wheels had sunk into the soft meadow soil and the fuselage was resting on the ground. The battered old helicopter with its oil-streaked hull looked too much like a death trap. Guido turned away from the distressing sight to study the weather. The gathering of low clouds forced his decision. He marched resolutely to the helicopter and squeezed himself in with the others. It would be a tragedy for the group and their invaluable data to perish on the return trip, but what choice did they have? The doors closed—barely. The passengers sat grimly along the sides of the fuselage, trapped by the mountain of gear around them.

Among them were elite fly fishermen and some of Russia's and America's top scientists. Together they had explored one of Russia's many untouched salmon rivers on the peninsula of Kamchatka. Somewhere in the chopper sprigs of vegetation and river water floated in tubes and canisters of liquid nitrogen preserved fish scales and fin clippings. Packed away in waterproof bags were notebooks that held sketches and topographical measurements. Together, these gathered bits described an immense ecosystem Americans hadn't known existed. It was an ecosystem based on salmon. If they made it home, the team of scientists could help to change the way people saw salmon and their rivers forever.

The difficulty was in reaching this remote region, and in getting back home again.

Guido settled himself among his companions. Churning at full throttle, the turbine could barely lift them. It groaned and trembled before heaving itself upward, then rose slowly above the forest, sending leaves and branches whirling. The pilot dropped the nose of the chopper and headed south, flying so low over the forest that the branches seemed to brush the bottom of the craft.

They stayed low for a hundred miles, flying over the Kolpakova, Varavskaya, and Kol rivers. Through the windows the passengers could glimpse a vast landscape of forests, rivers, and mountains. The Sea of Okhotsk glittered to the west. To the east were mountain slopes dusted with the first snow of winter. Below were hills blanketed in a gold-and-orange mosaic of stone birch and river valleys marked by still-green bands of cottonwood and alder.

Guido had a view of the cockpit. The pilot was on the left, the navigator on the right, and the mechanic in the middle. It boded well that all three had some gray hair, and likely were veteran Russian pilots who had fought in the Afghan war. The pilot was looking for a break in the peaks. Guido watched his stoic face in half-profile as he decided on a route and headed straight toward the base of the mountain range. As the land began to rise, the pilot held his course. The chopper seemed ready to kiss the earth when the air pressure from the blades meeting the ground pushed it aloft. A few hundred yards later they were heading straight back into the mountain again, this time barely missing a copse of pines.

No one spoke as they stair-stepped up the mountain. Each passenger was sealed in his own isolation. Not all of them could see what was happening, but they could feel it. The chopper was hopping lamely from forest to tundra to snowpack. As they climbed higher, the engines labored in the thinning air.

Guido watched through the cockpit as the volcanic ridge above them approached. The vegetation below had given way to rock, snow, and ice. When they broke through the clouds, Guido got a partial view of two jagged rocky peaks with a slight dip in between, defined by a long ridge of windblown snow. The final precipice rose before them like a wall.

At the top of the ridge was a thin cornice of snow, as sharp as a razor. The helicopter throbbed and whined as it reached the summit. Then, they were through.

The other side of the mountain gave way to a thousand feet of vertical cliff. When the ground beneath them disappeared, the helicopter tipped over the cornice and fell. As the Mi-8 rolled slightly to the left, the cargo began to slide. One of the American passengers stood up and yelled "Oh shit!" and was immediately restrained by the Russian next to him. Any further destabilization could put them into a roll.

As the helicopter fell, its roar and groan were replaced by silence. The men felt weightless as the rotors thumped against the gravity of their descent. Long seconds passed as the blades slowly caught purchase in the thin air and stabilized the chopper. Gradually, it moved forward as well as down, arcing out across the mountain range into the broad valley of the Kamchatka River and the city of Petropavlovsk beyond. Guido strained to see the rivers below. He recognized these transporters of fish as living organisms now, and wanted to observe the idiosyncrasies of their design, where their flow was pinched by canyons, where they curved back on themselves, and where they spread across the plains like light.

Of the many trips to Russia that followed, some would be to Moscow, to do battle behind the walls of the Kremlin. His battle would hurtle him into one of the most formidable, complicated, and unpredictable countries in the world. To protect these perfect rivers, he needed knowledge, money, and power. And he needed them in Russia.

Over the next twenty years and twenty-five expeditions, Guido would harness the power of oligarchs and billionaires on two continents to protect vast swaths of pristine wilderness and do more than any man alive to make the United States and Russia realize they were joint custodians of a massive watershed ecosystem held in place by salmon.

It was the salmon that had brought Guido to the fight. He had chased them for most of his life and seemed to take instruction from their wildness, their resilience, and their unwavering purpose. Perhaps the salmon, with their epic migrations and unflagging determination to live, inspired him to choose a course that even those closest to him could sometimes not believe.

It had all started long ago, on the other end of the Pacific Rim, with a wayward towheaded boy and his fishing rod on a river in Oregon. A boy who, from the start, was uncommon.

STRONGHOLD

CHAPTER 1

THE FIRST STRONGHOLD

THERE WERE TWO FISHING CABINS on the Deschutes River, nestled up against the Warm Springs Indian Reservation. Guido's family took a boat across the river and wheelbarrowed their provisions to their cabin. There were no roads on their side of the river, just miles of wilderness. The trails, mountains, canyons, draws, and creeks were part of their family mythology, a landscape both unknowable and intimate, like the river that flowed so full and strong even when they were not there, that changed every moment and changed not at all.

And cradled in this rough crèche, there were many ways to die. There were rattlesnakes coiled in the woodpile, scorpions sheltering under flat rocks, black widows spun into the corners of the boot room. None of them was as dangerous as the river itself; the river that had taken a beau of Guido's mother and a beloved cousin, swallowed them from sight, tugging them into the whirling currents and underwater vortices.

The other cabin belonged to our family. My father and Guido's mother were siblings, and in summers our families merged into one.

There were five children in the Rahr family, and three in ours. Clustered around the same age, we were knit close in that rugged country, playing games and making up stories that were too wild for most children.

Guido (pronounced *Gee-doe*) was the oldest of our pack of first cousins, and the only one who moved through our dangerous kingdom as if he belonged there. He had a natural grace and assurance and he rarely stepped wrong. As early as seven years old he would disappear for the day and rejoin us wordlessly, leaving us to wonder about the life he was leading, and how different it was from ours. He had been born with a natural intelligence that gave him a power we all recognized.

I tried to follow him. I was younger and a girl, but I was tough and I didn't complain. He tolerated me sometimes, but only if I moved through the landscape as he did, mimicking his silent walk, my senses alert. I stood by as he divined where a blue racer lay resting in the shade of a sagebrush, or when he lifted a rock that harbored a scorpion. He was no more afraid of a rattlesnake than a mouse, and his quick hands could catch anything.

He told me that even with his feet on the ground, he saw the earth from far above. I believed, in some shamanistic way, that he could. I believed that from his place in the sky, the stories of the land were apparent to him: those of rock formations, springs, volcanoes, and landslides. And within these large stories were smaller ones, each as complex and complete as a miniature solar system. He could make me see these things, and so was a kind of god to me.

I could not have guessed then how close we would one day be nor, indeed, that it was even possible to be close to such a boy.

I knew in a childlike way that Guido wasn't wired like most people. Nor was I. Since infancy, I had experienced brain tremors when for intense moments I trembled and shook and was unable to communicate. The episodes were both frightening and mysterious, for no doctor could explain why they happened. The tremors temporarily severed my connection to the world around me, and were terribly isolating. Guido was also isolated, but his was a strong isolation, and it was happy. Our slight maladaptations marked us as black sheep, the children the adults worried about most, the ones discussed in the hushed hours after dinner. But in those lonely, afflicted childhood years, Guido's contrariety was a gift to me. When I was with him, I lost track of time. His simple, focused world delivered me from the painful vicissitudes of my own. I didn't grasp then what an impossible child he was for his parents to raise—or how hurtful he was to the family he more or less ignored. My aunt Laurie later told me how she and Guido Sr. had struggled with a son who had no interest in learning what they had to teach him, or indeed in what the wider world considered a conventional education.

What Guido wanted, from infancy, was the freedom to pursue his own learning, unimpeded by the many adults who oversaw his life. Another world beckoned to him, and the wilder the better. When his parents denied him access to it, he ran away. By the age of four, Guido had run away multiple times to roam the fields and streams around their home in Oregon; the local police became familiar with the boy who vanished into Lake Oswego's undeveloped swaths of suburbia, where he captured anything that moved and befriended it long enough to study it. Back in his room, he filled notebook upon notebook with renderings of insects, reptiles, and fish.

His mother was struck by the accuracy of his drawings. In preschool his precocity was noted by his teacher when she asked the class to draw what they thought their insides looked like. Most of the children drew circles filled with scribbles. Guido's drawing had intestines and bones and organs. His teacher was so astonished she called his parents to inform them that their four-year-old had a rudimentary understanding of human biology. Laurie was mystified. The only way Guido could have known such things was by studying the family *World Book Encyclopedia,* and what four-year-old did that?

His first-grade teacher, Helga Peters, told Laurie that she'd never had a student like Guido; that in her experience most children were like sausages—you could stuff them full of knowledge. But Guido wasn't like that; he knew exactly what he wanted to learn—and it wasn't what she was teaching.

She wrote,

> Guido has been the most original student during his first year in our primary grades. The academic tools, i.e. learning to read, learning to write and coping with mathematical concepts are to Guido an evil to be suffered. . . . He wants to read and write about the mysteries of nature, the creatures and the glorious wonders of creation. Although he can express himself magnificently in all art forms he is trying hard to read and to write, not easy when he has to match his excellent vocabulary. Since his speech is fast and slurred it is very hard for him to sound and spell words.

Up until this point, his parents had no idea anything was unusual about their firstborn other than that he was spirited and independent. "We didn't know he was different," Laurie said. "It was other people who told us he wasn't normal." As Guido grew older, the Rahrs began to struggle with their expectations of him. "That he simply wouldn't cooperate came as a terrible shock to us," Laurie remembered. Guido Sr. was bewildered by his son's obstinacy and a character so unlike his own. The product of a genteel aristocratic German family, Guido Sr. was a pipe smoker and reader of poetry, and his love of nature ran to the romantic. He preferred fields of wildflowers to the high desert, the Alps to Oregon's rivers, a vintage rifle to a fishing rod. The rest of the Rahr children pleased Guido Sr. greatly; they were well behaved and seemed to understand the order of things. His firstborn continued to thwart his expectations.

One time the family set off for a posh tennis camp in Arizona to polish both their children's games and their etiquette. On their first day there, Guido failed to show up for a single tennis class. Somehow he managed to escape the compound and slink into the wilderness to see what lived in the bushes and under the rocks of the John Gardiner Tennis Ranch. He knew there were collared lizards and banded geckos. Of particular interest were the chuckwalla lizards, native to the region. These lizards did not live in Oregon, and Guido was eager to get his hands on one. When the family returned after their day of tennis, Guido was nowhere to be found. A scream issued from his mother in the bathroom when she found the bathtub filled with foot-long chuckwalla lizards.

"We couldn't keep track of him," his mother said. "He always had his own agenda—and we were not party to it. It was only afterwards, when it was a fait accompli, that we learned what he was up to."

While he showed an aptitude for natural science, it was clear that something else was going on with Guido. He had an aversion to reading, and to any book that didn't involve pictures of reptiles. Unlike his educated and literary parents, he did not enjoy school, nor did he excel in classes other than art. Midway through elementary school, Guido was diagnosed with dyslexia. Special tutors were called in and he was forcibly held to the studies he loathed. When he was released, he disappeared, melting into the woods like a spirit. In fourth grade, Guido showed a glimmer of a deeper capacity. On a hunch, his mother gave him Raymond Ditmars's *The Reptiles of North America*. It was a thick academic tome, filled with Latinate terms, scientific descriptions, and beautiful black-and-white plates and illustrations. "His eyes grew big," she remembered. "He sat down with it right away. I think it was the first book he had ever wanted to read." His parents watched as Guido bored into the pages of Ditmars's guide. "It took him weeks to get through it, but he memorized it cover to cover." Laurie Rahr was fascinated by Guido's determination. It seemed that by sheer will, her nine-year-old son overcame his dyslexia.

While words were cooperating, numbers were another matter. Numbers and the language of math did not make sense to him, and he tried hard to understand what seemed rudimentary to everyone else.

Guido was eleven when the Rahrs moved to Minnesota and his father took over the family malting business. Guido's life took a turn for the worse, for the wildlife here was paltry, and the reptiles were limited to painted turtles, garter snakes, and the occasional skink. Most of what he could catch was uninteresting to him. I visited then, and found Guido in a desperate state. His blond hair hung long and lank over blue eyes that had lost their spark.

To survive the confines of his room, he had created a wilderness. Snakes and lizards came and went from open cages and sunned on pillows and windowsills while Guido sat in the middle of the floor and sketched them. We set off in a canoe to scour the banks of Lake Minnetonka for creatures, but mostly we just wanted to be away from the manicured lawns and gardens of the Rahr property. All we found were ducks, which we chased until they flew away. It was still thrilling to be with Guido, and to be included in the hunt. Even with the ducks on Lake Minnetonka, I knew we were not playing a game. Hunting was a deadly serious pursuit that required every bit of our attention. I had to be aware of every stick and leaf, of stepping right and pausing to wait when Guido paused, and holding my breath and remaining motionless when he held up his hand. I don't know why it was so much fun, except that I had become something of a hunter too. Guido had activated the instinct in me, and he'd done it by collapsing the distance between me and the natural world.

It was only years later that I thought about how he'd done such a thing. It was partly his intensity and focus, for these were contagious. Mostly, though, I think it was his preternatural understanding of a realm that most could not see. One can sense expanded awareness in another person, even if one cannot understand it. With Guido, I became a part of the trees and wind, the buzz of insects and flitting of birds. I could sense the life in the undergrowth and beneath the surface of the water. There were few thoughts in this place, and for long moments my mind gave over to an overwhelming feeling of connection and a sensation I can only describe as joy.

The social nets that might have caught other children failed to hold Guido. Because of his after-school tutoring, he was excluded from sports teams. As time went on, sports and social events lost all appeal for him. He had no friends to speak of, which didn't bother him much, as snakes had become his preferred company. He spent his days thinking about them and the crisp intricacy of their design. Nothing he had seen in nature was quite as beautiful, or mysterious. While other children took notes, Guido sat in the back of class sketching lizards and snakes and the sagebrush of Oregon's high desert in the margins of his notebook, a diversion that did not result in good grades. While his mother admired his artistic ability, Guido's father was less enthralled. Guido Sr. was a conservative man who would have preferred that his firstborn son take over the family business, as he himself had done. His father recognized that the privilege he had bestowed upon his son might be wasted on him; more concerning was that his son eschewed opportunities, turned his back on open doors, and every year drifted further away from what was conventionally acceptable.

Guido continued to struggle in school, especially in math. His recalcitrant attitude did nothing to help. The faculty at the exclusive Blake School had not warmed to him—nor he to them. The one teacher Guido held out hope for was Richard Green (not his real name), a fellow herpetologist, who had set up a "live" room in the basement of the school that housed a veritable zoo of reptiles. Green kept the capacious room dark and warm. It was filled, floor to ceiling, with cages that held boa constrictors, pythons, an anaconda, and even a pair of Gila monsters. In large enclosures on the floor lounged nine or ten alligators and crocodiles, the largest being a two-hundred-pound American alligator that hissed in the darkness when Guido walked by.

The live room was an oasis for Guido—and Richard Green was a beacon. Anxious to make a good impression, Guido volunteered to help Green take care of his reptiles, changing their cages and feeding them while his teacher taught upstairs. Green had no objection to Guido's help, but when the boy attempted to engage him in conversation, Green made it clear that he was not interested in Guido, or his knowledge of reptiles. To Guido, who had few if any allies, it felt like a crushing betrayal.

The mutual distrust between Guido and his teachers grew as the boy slipped further from sight. He spent every possible moment in the live room, where it was clear to him that some of the animals were failing to thrive; a few grew so weak, they faded from life altogether. Guido did what he could to nurse them back to health and was, as a result, perpetually late for classes. His excuses invariably involved a complication with an ailing snake or a lizard who was off its feed. The faculty was tiring of Guido's delinquency and what they considered to be his colorful fabrications. One afternoon when Guido emerged from the darkness of the live room with dilated pupils, a teacher accused him of doing drugs. In the 1970s, use of drugs, particularly psychedelic drugs, was on the rise in the country, and Blake had adopted a hypervigilant stance. Ignoring Guido's insistence that he had never taken LSD or anything like it, the school called the Rahrs into the dean's office and told them that their son was doing hallucinogens.

Guido had barely recovered from this injustice when he was caught for something else. This time he was guilty. He and another boy had brought flasks of gin and apple juice to the school prom. They hadn't even tasted the questionable cocktail when they were apprehended by chaperones. Unlike the other boy, Guido was not simply lectured; he was suspended and sent to a youth treatment center for a week. It was, in the eyes of the Rahrs, an overreaction that further stigmatized their son. When he got back to school, the other kids steered clear of him.

Guido discovered he could pursue a superior education from home. By mail and phone, he befriended an older crowd who had knowledge of reptiles, many of whom had doctoral degrees. He soon learned how to buy, sell, and swap snakes by US post. His allowance and chore money were spent procuring rare species by mail and, unbeknownst to his family, the menagerie in his room grew. The adolescent boy was also becoming an authority. Members of his chosen tribe recognized both his intelligence and his knowledge, which was rapidly expanding.

By twelve, Guido was well versed in global reptile dealing. He was particularly keen on rare boa constrictors, specifically from the Caribbean. Endemic species from these islands were beautiful and hard

to come by, but he had located a dealer in Florida who had access to
such species, marvelous big snakes with exquisite markings. Guido
was on the phone with the dealer every week, peppering him with
questions while he raised money for a Caribbean boa. The dealer assumed Guido was an adult because he spoke with such authority and
referred to the snakes by their Latin names. He realized Guido was a
kid only when his mother picked up the phone one day, listened in for
a moment, and promptly shut down the conversation.

Increasingly, Guido existed on the fringes of his family, with little
to no interaction with his four younger siblings. "I'm not sure he was
even aware of us," his sister Sarah remembered, "or our pets—which
he had no respect for." It was true, Guido couldn't have cared less
about the family dogs, or his sister's guinea pig—these vulnerable,
warm-blooded creatures lacked the elegant design of reptiles and the
perfectly adapted physiques that allowed them to survive cold winters or months without food. Mammals, including humans, could not
elude detection or live for weeks under a rock. They could not swallow and digest live prey, wasting nothing. Mammals were born vulnerable, and from birth needed nurturing and care. It was only years
later that it occurred to me that Guido not only admired cold-blooded
creatures but identified with them. They were solitary, quiet, and
often misunderstood. They were also quick and crafty, and their survival depended on their ability to disappear from sight.

At fifteen, Guido moved off the grid entirely when he was offered a
job few his age could even attempt. His second cousin Spencer Beebe,
a budding conservationist, had recognized his younger cousin's gift
for catching and observing reptiles and recommended him to conduct
a survey on a vast tract of land that had been gifted to the Nature
Conservancy in central Oregon. Spencer insisted that if anyone would
be able to tell them what lived on the Lawrence Memorial Grassland
Preserve, it was this teenage kid.

Guido was given a three-week summer internship, and, with some
reservations, his parents dropped him in the middle of Oregon's high

desert to document snake and other reptile life. Here he lived happily in a rustic cabin and spent the days hunting for creatures, accountable to no one. When he returned to Minnesota, his parents could no longer ignore that Guido needed out, not just of Blake but of the Midwest. Far from his chosen habitat, he was failing to thrive. That year, they took him on a tour of reputable boarding schools on both the east and west coasts, but he instantly favored the one "alternative" school in Arizona. Verde Valley was composed of small adobe buildings and situated in gorgeous red-rock country. The school faced a monumental sandstone spire that dominated the horizon like a cathedral. As soon as Guido saw the vast acreage and high desert surrounding Verde Valley's campus, the tour was over for him. He promptly slipped away from his parents to explore the habitat of the mountain king snake, a lovely and mild-mannered snake with bright red bands that Guido had coveted for years. The Rahrs had a hard time getting their son back in the car.

It was the first time Guido had felt comfortable in an institution of learning. He was allowed to bring his snake collection with him, and he quickly added to it, taking full advantage of the school's relaxed atmosphere to live wild and skip classes to chase reptiles in the rim-rock of Arizona's high desert. On cultural exchanges, Guido was sent for a month to live on a Navajo Indian reservation, and later he went to a small Mexican village, where he practiced his Spanish working the fields with the locals. He felt perfectly at ease in these poor, rural communities with people who lived close to the land, a land that grew corn and wheat and was alive with birds and insects—a far cry from the concrete suburbs of Minneapolis.

After his first year, Guido came home with a footlocker of dirty clothes and lurked close by while his mother sat in the laundry room sorting through the pile. As she was nearing the bottom, Guido looked in and advised her to stop. Laurie rose and backed up, knowing what might be in store. From the bottom of the locker, Guido pulled a strangely weighted pillowcase and opened it for his mother, who took a quick peek. Coiled within it was a large black-tailed rattlesnake. By way of explanation all Guido offered was, "It was too beautiful to leave behind."

The Rahrs' cleaning woman did not fare as well. The poor woman was terrorized by Guido's wandering menagerie; her nerves were frayed after finding one too many snakes coiled in closets. Finally she snapped when she discovered a snake burrowed between sheets in a laundry basket, at which point she got in her car and drove away. Later she had her husband call and tell the Rahrs she was never coming back.

I saw Guido at our river cabins in the summers, and throughout the year I heard tales of his misbehavior. As a teenager, he was unkempt and hard to see, his face obscured by long hair and tinted glasses. He was bored by and unfit for social interaction, and he rarely joined us at mealtime. Having a normal conversation with him was pointless, and I never tried. It was by following him that I learned. Once in a while I proved useful to him by catching interesting reptiles. Sometimes, when warranted, he paused his own furious hunt to acknowledge me as someone with occasional value. These moments of recognition left me beaming, for Guido was fiercely real, as pure and unapologetic as nature itself. To register in his world, even momentarily, meant more than the approval of any adult.

At fifteen, Guido had become interested in catching the fish that swam in the Deschutes River. The rest of us were already fishing, catching and keeping the allowed twelve-inch rainbows that we called "breakfast fish," for that was when we fried them in butter, peeled the meat from their delicate bones, and ate them. Guido, who had been wholly preoccupied with reptiles, came late to the game. When he decided he wanted to learn how to fly-fish, however, our grandfather resisted teaching him, possibly for less than honorable reasons.

Our grandfather was a proud and competitive angler who had founded our little community on the Deschutes, building its rock walls and planting its great shade trees that grew leafy and tall, creating an oasis in the middle of the high desert. The Deschutes was his personal fiefdom, and he had named every hole, riffle, tail-out, and stump that yielded fish. This charming, mercurial man was the founder of our family stronghold, a place we sheltered from the outside world.

Grandfather did not yet know what to make of his first grandson, who roundly ignored his authority. The boy was an irritant in other ways too. While Grandfather claimed a great affinity with nature, it was clear that his grandson's relationship to the natural world went deeper, that somehow the boy's understanding and instinctual knowledge of the river trumped the old man's. It might have been pride or ego that prevented Grandpa from teaching Guido how to fly-fish, and jealously keeping the secrets of the river from him. Guido responded by committing the capital offense of bringing his Minnesota spinning rod to the Deschutes, where he proceeded to hook and catch the prize fish of the river.

The distinction between fly-fishing and regular fishing is, for some fly fishermen, as fraught as a dispute between warring religious sects. In our family dogma (laid out by my grandfather), fly fishermen were more respectful, intelligent, and even morally superior to lure fishermen. It was the difference between a hunter who stalks his prey for days versus the weekend warrior who shoots from the back of a truck. One is fighting fair, and the other is a kind of terrorist.

The virtues of the fly fisherman run deep. He meets a fish on its own ground, casting a fly attached to a tiny barbless hook (so as to not injure the fish with its removal). This fly has been tied of tiny strands of feather and fur to make it imitate the flies the fish feed on naturally. Casting such a fly is no easy matter, for it must be made to land on the water like any other fly. Getting such a light object to travel through the air with speed and direction requires momentum. This is provided by the angler, who casts his line forcefully forward and back before sending it sailing across the river. The backcast is an ongoing trial that, if survived, can bestow the angler with the precious gift of patience. The banks of the Deschutes were annoyingly filled with vegetation, and as children we spent as much time untangling our flies from branches and trees as we did fishing. The knots we so painstakingly tied often did not hold, and our flies were lost to the sagebrush and oaks. I have memories of screaming in frustration.

The fly fisherman often wades into the river to get away from trees, or to access a certain patch of water. Here, he becomes part of the river and stands often for hours with the current tugging him. Navigating the slippery rocks he cannot see, he casts his fly, fixing an eye on it as it floats downstream. Then, with every sense available to him, he becomes attuned to the exact moment the fly is taken by a fish from below and disappears from view—at which point he raises the rod to set the hook. It is one point among many at which the fish may escape.

The reasons for success or failure in fly-fishing are multitudinous. It is a challenging, if not maddening, sport, and one with no guaranteed reward. There are many mysteries as to why a fish does or doesn't take a fisherman's fly, and only some of them are linked to the angler's skill. These mysteries are mulled over obsessively by anglers like my grandfather, for whom fish and parts of the river assumed almost supernatural powers, inspiring him to grapple with nature's complexity and accept (as well as he could) his own mortal limitations.

A tackle fisherman (also called a spinning, bait, lure, or gear fisherman) has an easier go of it. Standing on the bank, he casts a weighted lure and lets it sink straight down into the water, where it may very well bonk a fish on the head and be snapped at out of irritation. If ignored, the lure is then rapidly reeled in and travels through the water with enticing speed, flashing brightly in the darkness. Compared to a homely dry fly, spinning lures are tantalizingly perfect imitations of minnows, and they prey on the vulnerability of surprise as well as a fish's sensitivity to light. Once swallowed, the two treble hooks, one attached at the middle and the other to the end of the lure, are not designed for easy removal.

The big trout of our home water on the Deschutes went for Guido's glinting, diving Rapala lure, one after another. These were important fish we all knew about, the four- and five-star generals—big resident rainbows that often broke Grandfather's line as many times, earning them their stars. Guido caught many of them, and he did so easily and, to our grandfather's horror, *barbarically*—with the insult of a spinning rod. To Guido it didn't matter; he was going to catch fish one way or another. He landed them, netted them, hooked his fingers through their gills, and paraded them past Grandfather's cabin without a sideways glance. It was a declaration of war. Helpless and enraged, Grandfather called an emergency meeting with the aunts and uncles. The question was what to do about Guido; the boy had to be stopped. What he was doing was indecent—it was criminal! They decided that the only thing to do was to teach him to fly-fish.

We shared the Deschutes with the Warm Springs Indian Tribe, whom we rarely saw because they lived on the other end of their large reservation. Like other Native American tribes in the Pacific Northwest, the Warm Springs Tribe had exclusive rights to fish for salmon, and the tribe did so mostly twenty miles or so downriver, at a fifteen-foot waterfall where the salmon were relatively easy to net as they attempted to hurl themselves up the falls. So as children, we knew little of salmon; our domain consisted mainly of rainbow trout, a crafty and mysterious creature that often eluded our flies. Guido alone seemed to have the capacity to imagine their lives in the river—to follow them to a world that consumed him completely. Driven by a feverish determination to understand them, he made a comprehensive study not just of the fish but of their river. When Guido was fifteen, the Deschutes came bursting into life. I watched him lie in the sagebrush on his belly noting the insect hatches, the trees and the angle of their branches, the pitch of the bank and the shadows he cast, the bubbling currents and the still waters. And as he watched, I knew he was beginning to anticipate where the fish might be, and what they might be doing—or about to do. The end result was always capture, the proof he had understood it all right. The rigor of fly-fishing suited him perfectly. The harder he had to work for his quarry, the more he respected it.

After catching more than his share of four- and five-star generals the honorable way, Guido graduated quietly from trout to their grand seagoing relatives, steelhead. We rarely saw these majestic fish. They were big and powerful and swam in fast from the ocean, hell-bent on getting to their birth streams to spawn. They were nearly impossible to catch. As Guido conjured their lives deep in the current, he experimented with flies that sank to the bottom and swam slowly, mimicking the nymph form of stone flies that steelhead feed on. To the consternation of our elders, he got it right. Soon these glinting chrome-sided steelhead were flipping out of the water like little porpoises at the end of Guido's line. One day, when we were wandering up Eagle Creek, Guido gave me a proper introduction to this fish I barely understood. He took us off trail so we could stay close to the stream. Guido saw the steelhead long before I did and motioned me silently to the ground. For the next hour we lay still in the tall grass watching a steelhead hen fighting her way up Eagle Creek, her strong back breaching in the sun as she leaped up the last rapids—all the way back to the place where she'd been born. We inched forward to find that she had chosen her redd just yards away from where we lay. We could hear her scuttling stones and pebbles with her tail to make the nest that would protect her eggs. By the time she had finished her thrashing, her tail was ragged and torn, and her energy spent. Nearby were vying males, chasing off one another and returning to the shivering female, waiting for her to release her burden of eggs.

"She started out as a little rainbow trout," Guido whispered by way of explanation. I still didn't understand the connection between rainbow trout and steelhead. I couldn't see the relationship between this enormous green-backed fish and the lively little rainbows we caught. Guido shared another perplexing observation by pointing out that the primary contenders for the hen's eggs now were a steelhead buck and a resident trout. The trout was small, dark, and opportunistic. Guido explained how he had opted to play it safe in the river and was now edging in on the action of a buck who had traveled thousands of miles to reach this place, this moment. The buck, with his deep red sides, wearily chased the pesky trout away with the last of his energy. The truth of the matter, Guido added, was that the nobler fish did not always win the prize. Either fish could inseminate the hen's eggs. This seemed wildly unfair to me, but Guido accepted such realities without judgment or sentiment. Very early on he understood that nature had its ways, and there was a reason for everything.

We would return to this spot in the late spring and early summer because Eagle Creek ran dry by July, and minnows could get trapped. Grandfather had started a tradition that we dutifully carried out every year, hauling buckets and nets up the canyon to rescue the tiny fish in their isolated pools, scooping up the darting schools that gasped for oxygen in the warming creek. We released them in the calm inlet where Eagle Creek met the Deschutes. Here they quickly revived in the cold water, gathering together like a little army to face the swift current.

As Guido spent more and more time in the Deschutes, his relationship with steelhead intensified. This fish seemed to have taken the place of snakes and lizards, perhaps because it had more to teach him. Steelhead were every bit as enigmatic as he was, and every year during the brief steelhead season, it seemed the two of them conducted a long, private conversation at opposite ends of a fishing rod.

I came to regard the fish swimming in our river as shape-shifters. The rainbow trout of the Deschutes could, under certain circumstances, transform their physiology and become anadromous: able to live in both fresh and salt water. They could leave the river as rainbow trout and come back as salmon. I still find it baffling that a creature can start life as one creature and end as another, like a dog going to the woods for a year and emerging as a wolf.

When they return to the river from the sea, steelhead are long, silver, sleek as bullets, and look nothing like the jewel-toned rainbows they once were. They have become phenomenal swimmers, and have traveled thousands of miles. When you look at them, you have no question that they have joined the ranks of Pacific sea-run fish, the great tribe *Oncorhynchus*. Some fish scientists would argue that they have actually become salmon. Some say this isn't possible. There are places where the human ability to understand the natural world falters.

For all their likeness, steelhead don't really behave like salmon. Salmon have been swimming the earth for millennia and are tough, resourceful fish. When the melting of the ice caps gave them access to the rich offerings of the sea, they adapted to salt water, transforming their lungs and kidneys to process this new element. When they are strong enough, young salmon leave the refuge of their birth rivers and swim downstream for the ocean, sheltering briefly in briny estuaries as their physiologies change. They then disappear into the ocean, swimming as far as Japan in their hunt for food, and staying away for up to four years. When it is time for them to spawn, salmon stop eating and swim all the way back to their birth rivers, repeating the journey in reverse. It is an epic effort that, after spawning, ends abruptly in death. The scientific term is *semelparity*—a dramatic strategy that makes biological sense, for the likelihood of a salmon surviving such a grueling journey more than once is slim. So they give it their all, just once.

But steelhead deviate from this pattern, and others. While salmon travel to sea in schools, steelhead strike out on their own. They don't turn right and migrate up the coast as other salmon do, but head straight for the open water by themselves. Here they might meet up and travel with other steelhead, swimming fast, covering up to ten miles a day toward whatever waters are calling them. Then they stay for a year, or two—or four. When they return to the river, they don't die after spawning. Possessed of seemingly supernatural strength, they head back out to sea and return to spawn, sometimes multiple times.

One would think that such a strenuous program would end in early death, but steelhead are one of the longest-living salmonid in the Pacific, with life spans of up to eight years. They are also the rarest.

How is such a creature possible? As I write this, no one can explain beyond vagaries how steelhead came to be—or the process involved when a rainbow trout decides to leave the river for the sea. It seems to be an individual choice, if one could endow a fish with decision-making agency. There is seemingly little, if any, genetic logic involved. Two rainbow trout can spawn a steelhead just as two steelhead can spawn a rainbow trout. Whether a trout goes to sea or stays in the river is determined, no doubt, by many micro variations.

I like to think that maybe it is simply the spirit of the fish. Maybe, like some people, some trout strike out for the big city to try their luck at fortune and adventure, leaving their home rivers for the wide Pacific, where food is rich and they can grow bigger than they ever would in the river. The bigger the fish, the more eggs it can carry—and more eggs equals success, for it increases the probability that one's offspring will survive. A rainbow trout lays eight hundred eggs while a steelhead lays three thousand. Such a payoff is very likely worth the long and dangerous journey, a journey that brings other transformations. In their bid for biological advantage, rainbow trout shed their troutiness and become steelhead, joining perhaps the most spectacular fish on earth—salmon. So similar are they, that some biologists consider them to be the seventh species of Pacific salmon.

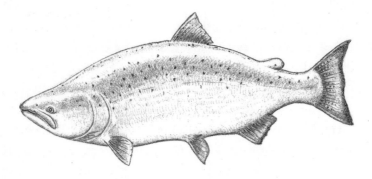

CHINOOK SALMON

It's unlikely that one finds oneself falling in love with a fish. But in this, I followed my cousin once again. With salmon, Guido found a creature that captivated him completely. These fish were the purest expression of life he'd encountered, and seemed to hold answers to questions he had yet to ask.

When Guido described the six species of Pacific salmon to me, I saw them clearly. There were the leaping, whimsical coho, who played and worked as a group; chinook, the kings, who were fiercely strong and organized, traveling far and wide in deepwater schools;

sockeye, an enigmatic lake species who fed daintily on insects and zooplankton in slow water and turned from silvery to ruby red when spawning; the tough but less distinguished chum, who migrated at just a few weeks old in an undulating silvery cloud; and the small, plentiful pinks, whose range stretched all the way to Japan and who returned like clockwork every two years to their birth rivers. Cherry, or masu salmon, were a rare Asian species that turned bright red when spawning and seemed almost like trout. Lastly were steelhead, the possible seventh species, who were quick, watchful, and solitary, and could hold in the fastest water.

I have since wondered why Atlantic salmon did not evolve this way and remained, instead, a single species. Were they simply perfect as they were? Or did it have to do with something environmental— perhaps they were stymied by the colder water of the Atlantic, or uninspired by the shorter migration to their feeding grounds? Or maybe it had to do with the older, more sedate geography of the North American plate. Maybe the color and diversity of Pacific salmon simply mirrored the drama of their coast, where the grinding Pacific plate still pushes the earth's magma into soaring mountain ranges that host surging rivers and lakes. Maybe Pacific salmon adapted to inhabit the myriad niches of the newest geography on earth. Maybe salmon, like humans, reflect the terrain to which they are born. That Atlantic salmon have all but disappeared from both Europe and the eastern seaboard makes Pacific salmon even more precious.

Each of the six species of Pacific salmon has a "life history," or a relatively fixed behavioral course. They spawn in different places of the river, laying their eggs in different kinds of water and in different-sized gravel; they eat different foods in different parts of the ocean; they mature and migrate at different times, staying in the ocean for different lengths of time. Staggered in this way, they do not compete, rather they are complements, colorful pieces to a mosaic that de-scribes one of the most stable and productive ecosystems on earth.

For whatever reason, the diversity of Pacific salmon has been key to their abundance. With the ever-changing conditions on earth,

where no two years are the same, one species will prevail over another. For instance, in one year the ocean currents, temperature, and
rainfall will benefit chum over chinook. The next year it will change.
The end result is that no matter how tough the conditions, some
salmon make it home to spawn.

Every natural-born salmon has a home, a river or other body of
water to which it belongs. And it has a place within that river, a bend
or a gravel bar or a pool that is as familiar and safe as a hometown,
with just the right depth and flow, and the right-sized gravel in which
to lay their eggs. A salmon's fidelity to its home river, to the exact
patch where it was conceived, defies our modern perception of the
randomness of nature, but then very little in nature is actually random. There are patterns, if we can but see them. But salmon disappear from our view when they enter the sea and are, at best, half
known to us. We can watch them only at the beginning and end of
their lives. What happens in between is a mystery. With electronic
tracking we are able to see more and more, but there are major questions we are not close to answering. For instance, how do they find
their way home from two thousand miles away? It is said that salmon
navigate by the sun and the moon, that they are aligned with mag-
netic north, that they can sense in the cubic masses of salt water a tiny
droplet from their home river. Perhaps all of this is true. What is in-
contestable is that salmon are some of the most determined, resilient
creatures on earth. For hundreds of thousands of years nothing has
prevented them from reaching their birth rivers to spawn—no matter
the conditions or obstacles. In the span of our lives, this seemingly
unalterable reality would change, changing Guido's life with it.

In my lifelong study of my cousin, I would return again and again
to these fish he chased, and ultimately devoted his life to. I watched
their struggle become his. I watched as, over time, their spirits merged.
In Native American lore, salmon people are guided by an inner knowing, and an understanding that some bends in the river bring ease and
others hardship, but all are valuable in what they teach. Salmon people understand that hardship is learning, and that what is started
must be finished. Where institutions failed Guido, nature did not. I

think it's possible that he took his deepest lessons from steelhead, and the tribe *Oncorhynchus,* the salmon. I think he was instructed by their majesty, strength, and singularity of purpose, and by their ability to adapt, to change in their very cells. I saw this in the course of Guido's remarkable life; I saw that, given the right impetus, astonishing transformations are possible.

CHAPTER 2

TUMBLING DOWNSTREAM

THE VERDE VALLEY SCHOOL in Sedona, Arizona, was a bend in
the river that brought ease for the first time in Guido's academic career. The school was filled with the sons and daughters of
musicians, movie stars, and children like Guido, who needed a little
more educational "flexibility." The leniency was fully exploited by
some students, Guido included. Along with the treasures of the So-
noran Desert, he discovered beer, Marlboro cigarettes, and the joys of
topless coed sunbathing. While even the most liberal schools had academic requirements, Guido found he could ignore them and instead
stage a years-long exploration of the remote canyons and rim plateaus of the Mogollon Rim, which he considered his true curriculum.
In his senior year, he rarely attended class. Instead he pursued mountain king snakes and rattlers in the red dust of the desert.

He passed his classes doing the minimum of work, unaware that he
had pushed the system too far. When it came time for graduation,
Guido was faced with his comeuppance. The school administrators
made the unusual decision of electing to prevent Guido from sliding
through with the rest of his class. They telephoned the Rahrs in Min-

nesota and advised them to accept their strong recommendation to withhold a diploma; Guido was a truant who had no regard for the system. It would be a mistake to let him proceed at this juncture. He needed to take his knocks in the real world. While disappointed, his parents weren't entirely surprised. Reluctantly, they agreed.

Guido accepted his sentence stoically. The filament of trust remaining between him and authority had disintegrated, and that was that. It didn't take much deliberation to decide on his next move; he would return to the river. Releasing some of his snakes into the desert and bagging up the rest, he set off for Bend, Oregon. He'd heard a rumor that lurking somewhere in the deep pools of the upper Deschutes River was a secret population of unusually large brown trout.

Finding and catching "trophy" trout was a momentous challenge in the fly-fishing world, and big browns were the toughest of all trout. Only a few elite anglers knew how to catch them, and these men did not share their secrets. There was nothing for Guido to do but teach himself, which suited him fine. He left without his parents' blessing or financial support. If he was going to fish for a living, he would have to make his own way—and he would have to start at the bottom of the food chain.

Bend in the seventies was a medium-sized town surrounded by high desert, alfalfa fields, and magnificent snow-covered mountains. Guido didn't know a soul there, and he had little money. These things mattered not at all; the air was pure and clear, and the upper Deschutes River flowed cheerfully through the center of town. Work for young drifters like Guido was scarce, but living was cheap. He found a room in a low-rent group house, got a job at Shakey's Pizza, and promptly began a methodical pursuit of the browns. Every free day he spent alone in the canyon land of the upper Deschutes, surrounded by basalt cliffs and springs fringed with watercress. He stood for hours in the deep pools among the cattails, junipers, and ponderosas, intoxicated by the ambrosial smell of sage and looking for the browns, but they were nowhere to be found.

At the local fly-fishing shop, the Fly Box, Guido met a fishing guide named Chip Chipalla, who knew the Deschutes like the back of his hand. Chip was a loner like Guido, with a scraggly beard, a cowboy hat, and an old pickup truck. He told Guido he had an idea of where the big browns might be hiding. Soon the two of them were jouncing off road in Chip's truck looking for pools they thought could be holding their quarry. Chip had a spot in mind far upriver. It was getting late when Chip pulled the truck over and led Guido to an overlook. Below was a stretch of deep, slow water that they could see clearly from the high rocks above. As they stared down, the wind stirred the scent of sagebrush and dimpled the pool with tiny ripples. Moments later, a sizable brown back rose from the depths, cresting and rolling through the water before disappearing from sight. The two men stopped breathing—then they whooped.

The next day, Guido was back at the pool on his own. Sitting under a juniper tree, he watched the water from above. It was getting toward dusk and the cliff swallows had started to feed, tilting and angling their fine sharp wings as they caught insects in midair—the same insects fish fed on. Guido sat up when something disturbed the water below, sending rings across the pool. Soon the biggest trout he had ever seen broke the surface and rolled, giving Guido a long look at its green back and orange sides dappled with black, dime-sized spots. This was his pool, he thought; this was his fish. He scrambled down the rocks to the bank, where he could see the browns rise up close. The water churned and roiled as the fish rose to aggressively feed on what Guido could clearly see were caddis flies, large aquatic moths.

Guido knew to offer a fish something it was hungry for, but he started fishing for the browns using the flies he had, thinking he might get lucky. He had caught his share of big, wise fish, and he could cast a pretty-near-perfect fly; nine times out of ten it fell gracefully from his leader and alighted on the water as if blown there by the wind. It went on to float along the current as if it were entirely alone; there were no telltale loops or kinks in the fishing line, no drag by the leader creating a tiny V wake, rendering it unnatural. But the old generals had seen such tricks before, and some had fallen for them in the past. They were wary of being tricked again.

A few days of perfect casts and Guido had not had a single bite. These browns clearly knew artificial lure from real food, and feather from fur. They were beyond selective in their feeding—they were maddening. Guido finally surrendered, got in his truck, and drove to the fly shop in town, where he spent money he didn't have on a handful of man-made caddis flies. Then he drove straight back to the pool.

In the following days, he haunted the riffles and holds that harbored the browns—casting his line again and again, without luck. When he wasn't on the river he thought about being there. At Shakey's he absentmindedly overdressed the pizzas, dumping double the cheese and sauce and whatever else had been ordered. The toppings of his pizzas bubbled over the sides of the dough, igniting flames and smoking in the oven. The manager warned him once and then again, but Guido was lost in thought trying to figure out what he was doing wrong with the browns. He had still not caught a single fish.

He decided to back off from fishing and instead simply watch the browns. There was clearly something he had missed. On sunny days, he paced the banks of the river with a pair of binoculars and polarized sunglasses, scanning the water, allowing his eyes to learn the vague forms of the browns against the dappled bottom of the river. When he saw a fish, he marked the spot in his mind. Gradually he began to learn their habits. The browns did not like the light. They moved out of the deep, dark water in the evening, after the sun had set. Guido camped out with his sleeping bag and tried, unsuccessfully, to fish for them at night. It was in the morning, when the first rays of sun hit the water, that he caught sight of their large, dark shapes holding in the clear water below. Then, as the day broke, he watched them retreat back into the shadows. Slowly, their world revealed itself to him.

The problem, he realized, had to be with his fly. In general, he had not paid much attention to flies. For years he'd been using standard flies purchased at fishing shops. But he was becoming convinced that the fly shops didn't sell what he needed for these fish. He took a close look at the flies in his box. Then he looked at them under a magnifying glass. They had been constructed by fly-tying experts to imitate an insect naturally occurring in the river. A few artfully combined strands of fur and feathers were wound tightly together onto the end of a hook. But it was a precise science. A basic mayfly, for instance, had to look and behave like all the other mayflies that were hatching above the water. Guido had studied them often. After tunneling out of their burrows in the riverbed and swimming to the water's surface, mayflies cast off their larval casings, inflated their wings, and took to the air. Trout rose and held just below the surface, swallowing the flies from the frenzied cloud that landed on the water. Guido took out a mayfly and held it between his fingertips. How hard could it be to tie one himself?

In the next weeks Guido devoted himself to the art of fly tying, an infuriating, surgical process that he instantly loved. He knew from the moment he tied his first fly that he would be good at it; his artist's eye for detail and his ability to make sharp scientific observations were an intersection of his natural talents.

He practiced tying all kinds of flies, even midges, the tiniest of dry flies—tiny enough for two cherished goldfish swimming in a bowl belonging to one of his housemates. The next time the owner cleaned out the fishbowl, she was mystified by a tiny hook nestled in the gravel.

When he was ready, Guido brought his full attention to his first caddis fly. He sat at his tying table with its vise, feathers, and fur, strands of the natural world that he would bind together to float, sink, and move like something in the wild. It had to be convincing from every angle. He had learned that to tie a great fly it was necessary to see from the eyes of a great fish. The process was totally absorbing, for it brought him not just close to the river, but into it.

He was extremely pleased with his rendition of a caddis fly, and was truly baffled when the browns weren't interested. There was still something he was missing. He had tinkered with every variable he knew. One day he leaned his rod against a juniper, went down to the river, and sat to watch. When the fish finally rose, he thought he saw something he hadn't noticed before. It was hard to tell, but it was possible that the browns might not actually be feeding on the moths; they might be going for the pupa. These pupa were hatching from cocoons at the bottom of the river, where they were propelled to the surface by a bubble of air. It seemed to Guido that the fish might be snatching the pupa just before the bubble reached the surface, in the moment they broke out of their exoskeletons and flew off as full moths. Guido packed up his gear and sped home to his tying table. That night he created his own caddis pupa, fashioning the body from the fine fur from the back of a rabbit, the pupa's emerging wings out of a small piece of duck feather, and the legs with a wrap of rough grouse breast.

The next day, Guido woke at dawn and headed to the river, his new flies nestled in a patch of sheep fur on his fishing vest. Standing at the water's edge, he realized he had yet to glean how to make his new caddis pupa fly *behave* like a caddis pupa—to float underwater and rise slowly to the surface. First, he dropped the fly in the water and rubbed it into the mud with his boot to soften the feathers. Then he waded into the river and started experimenting. His first cast was short, and he stripped the fly (methodically pulling the line in so it didn't coil or drag in the water, potentially spooking the fish) to his feet before flipping it back into the air for another cast. He directed his next cast upstream, into the main flow of the river. This time he let the fly drift three or so feet before he felt it first tick bottom. Then, as the fly drifted past him, he raised his rod tip to create tension so that the fly was pushed to the surface, imitating the real thing.

Guido's caddis pupa had just begun to swing across the river bottom when it was stopped cold. Then his rod lunged forward and the reel buzzed as a fish started taking out the line in slow pulls, like a freight train. Only a big brown trout could fight so hard and deep, digging and thrashing across the pool. When Guido worked the fish close he could see he was right. In the same moment the brown saw him too, and with a defiant splash surged down to the tail end of the pool. From there, the fish swam hard downriver. Guido plunged into the river after it, swimming with his rod in the air. When the brown dove under some boulders, Guido dunked his rod under the same boulders so the line wouldn't break. It was a clever old fish, but it was not going to beat him.

Later, there were tales of the pursuit by those along the river who had witnessed Guido bobbing downstream, holding on to his hat with one hand and his rod with the other. It was in a pool downriver that the brown finally showed signs of fatigue. Dripping wet, Guido got his footing and coaxed it into the shallows, where after one last argument, the brown allowed himself to be landed. It was a magnificent fish, and Guido took his time inspecting every detail. The trout was heavy in the shoulders, long in the jaw, and a full twenty-five inches. His tail was wider than the spread of Guido's fingers, and his flanks were an iridescent golden brown with black spots encircled with glowing red. Beside himself with pride, Guido froze the fish in a block of ice for display, as his grandfather had done when he finally caught "Methuselah," a legendary Deschutes rainbow that had eluded capture for years. Guido drove his frozen fish all the way to Deschutes to show his grandfather that he too was becoming a master.

Guido later discovered that his new cast was called the Leisenring Lift. It was a technique established by angler Jim Leisenring on Brodhead Creek, Pennsylvania, in the 1940s, but Guido had discovered it on his own. It was a formative triumph. From this point on, Guido would no longer seek or accept guidance from other sources, guides, or fishermen. He was too fascinated by the repeated trial and error required to build a detailed picture of an ecosystem. By the time Guido had caught the brown, he understood things most fishermen didn't. What was more, this knowledge belonged to him and him alone.

As his life's purpose became clear, Guido would adapt this approach to other fish and ecosystems and, finally, to people, the most challenging species of all. To save fish, Guido realized, he would need to understand humans as completely as he understood the big browns of the Deschutes. Deciphering their behavior, customs, and organizing principles would become the ultimate test of his hunter's instincts.

In Bend, Guido had fallen in with a questionable crowd, characters who devoted their lives to odd jobs, drinking beer, and growing marijuana. One of Guido's friends grew pot plants in the woods, where she tended to them in the nude. One day she was discovered by a police spotter plane and was arrested after fleeing naked into the forest. More dangerous was the Shakey's gang, who drank too much beer and caused minor disturbances after hours. Low-rent living and alcohol erupted regularly into violence, and Guido got into the habit of piling into the street with the rest of his friends when they heard the cry "Fight!"

As the months passed, Guido became aware that he was adapting to his new environs. He noticed that his private-school accent was relaxing. He traded "isn't" for "ain't" and started dropping his g's. He welcomed this adaptation, for he saw no advantage to the trappings of privilege. Thus far, the socioeconomics of his family had neither interested nor benefited him, and he shed this faint patina of wealth like a snake sheds its skin. In Bend, no one could accuse him of being coddled. He was part of a different class now, and he belonged to a different world. This world, however, knew nothing about Guido. They would never have guessed that his family owned one of the biggest malting companies in the world—that the beer they drank was likely brewed with Rahr malt.

His parents despaired at Guido's intermittent fishing reports, and his pleas for money. They did not understand the import of his undertaking with brown trout; that he had meticulously cracked the fish's wild code brought them little comfort. The only time they heard from him was when he needed money, which he was increasingly clever at extracting. "He was becoming a con artist," his mother remembered, "getting us to support things we probably shouldn't have." What was worse was that he no longer confided in them. His parents, like other adults, were not to be trusted. "We were outsiders to him," Laurie said. Guido Sr. despaired. His grandfather, however, was highly entertained. He sent Laurie the letters Guido wrote to him trying to finagle money. When he inquired about why his grandson didn't simply ask his parents for money, Guido replied that his parents were "hard up." This obvious fallacy was hilarious to the older man, who had developed a respect for the boy's moxie.

Guido was fired from Shakey's and got a job at Skippers Seafood and Chowder, where he was required to yell out "hot chowder" when delivering soup in the restaurant. It was a humiliating practice, possibly meant to add to (or distract from) the authenticity of a seafood restaurant in the middle of the desert. This was trout country, not the seaside, and Guido remained obdurately silent while walking his orders through the restaurant, thus ending his brief career as a seafood waiter.

It was another "desert rat" who gave Guido the break he needed, a man Guido instantly recognized as a member of his own tribe. Donnie Kerr was twenty years older and worked with the creatures of the high desert, taking students out for days at a time and bringing this hidden world alive for them. He had a special relationship with birds of prey, and these wild raptors perched on his arm while Kerr introduced them to small groups of wide-eyed kids. Kerr had raised the money to build a museum outside Bend, where he hoped people could have an in-depth, interactive experience of the desert ecosystem. When construction for the High Desert Museum began in the early eighties, Guido was there to help.

Guido wanted whatever employment Kerr could offer. A frozen part of himself began to thaw at the idea of doing work that meant something to him, and by being employed by someone he respected. He landed the next-to-lowest-level job at the museum—overseeing a crew of workers who were part of the government's Comprehensive Employment and Training Act. These particular CETA workers were rehabilitated prisoners tasked with gathering rocks to build the museum walls. Guido thought it was a fine place to start. For the next weeks, the eighteen-year-old dropout and eight ex-cons spent long, hot days scouring the ponderosa pine forest for sizable basalt rocks. It was during these grueling days that Guido made an unexpected discovery about himself.

To break the monotony, he had taken to describing features of the high desert while the CETA crew toiled in the sun. At first the men ignored the talkative long-haired kid. When Guido started catching reptiles and snakes to illustrate his teachings, however, he got their complete attention. He fearlessly handled rattlesnakes and scorpions as he described their behaviors and adaptations. After the novelty of it wore off, the CETA workers became genuinely curious. What did snakes eat? How did they swallow and digest a whole rodent? Why did they sun themselves on the rocks? How did hibernation work? Why did they flicker their tongues? Guido surprised himself with a narrative skill he hadn't known he'd possessed. He could see that his explanations transformed the desert for some of the CETA crew; they were noticing things they hadn't before, and they wanted to know more. Guido realized that he might have more to offer the museum than physical labor.

Later, when the High Desert Museum opened its doors, Kerr hired naturalists to deliver daily talks to the public. Kerr himself gave presentations on raptors, offering visitors thrilling contact with hawks and eagles. Guido took a chance and asked Kerr if he could give similar presentations on snakes and reptiles. Kerr knew something of Guido's expertise, and after watching him catch and handle the reptiles on exhibit, he granted Guido a provisional spot.

Guido tried out his new role one quiet early afternoon. He positioned himself on a stool in the museum's presentation area with snakes curled up in his chest pocket and lizards clinging to his shirt. When museum visitors eyed his long hair and tinted glasses uncertainly, Guido sat up and invited them over. Did they want to hold a blue-belly lizard or a bull snake? Would they like to learn more about the creatures of the high desert? Once he had gathered a small crowd, Guido began to talk, sharing his own process of discovery, for he had visceral memories of his own delight at uncovering the lives of these secretive beings. His audience leaned in toward him, for his comfort with these unfamiliar creatures inspired confidence. Guido lost all self-consciousness as they grew more animated and curious. Soon he had them cautiously reaching out to handle the reptiles themselves.

By the end of the presentation, the barrier between Guido and the group of strangers had fallen away. In its place was a small but unmistakable bond. No one was more surprised than Guido, who was long used to being judged, misunderstood, and mistrusted. But here, in the High Desert Museum, among the snakes and lizards, he had emboldened strangers to take a journey with him. Without thinking, he had led his audience to a place they'd never been. And they seemed to open their eyes and come to life.

Kerr couldn't have predicted the hypnotic effect the boy would have on a crowd—or how perfect he was for the museum. Guido's strange, intense charm drew people from all over. Twice a day in front of a growing circle, he handled his reptiles, calmly telling their stories as if they were characters in a book. As he spoke, lizards perched on his shoulders and snakes wound themselves around his arms like Egyptian bracelets. As the months passed, Guido grew confident with his newly realized gift, and the identity that came with it. He was a storyteller, a whisperer, a bridge between realms. People who watched him left the museum with a greater understanding of the intricate life that sped by outside their car windows.

CHAPTER 3

THE BRINY ESTUARY

L AURIE RAHR WAS HEARTENED that her son was out of the restaurant business, but collecting rocks and talking about reptiles was still far from solid ground. She decided that her wildest
cousin, Stoddard Malarkey, might bring Guido around to the possibilities that lay beyond the High Desert Museum. Cousin Stoddy was
something of a myth in our family. Turbulent, boisterous, and a bit of
a scofflaw, he had taken his powerful body and his head full of brains
and become a professional boxer. After a decade of carousing, he had
settled down, married, and earned a PhD in English literature. He
was now a professor at the University of Oregon, where he taught
Chaucer and Shakespeare to graduate students. Like Guido, Stoddy
had flouted convention—and he had come out the other end. Also, he
was close by, teaching at a university Guido had a shot at getting
into. So, at his mother's urging, Guido hit the road for Eugene, bringing along one of his CETA friends for company, a Vietnam vet with
serious PTSD and wandering eyes who had a habit of muttering to
himself.

Stoddy met the curious pair at the door, took in their odor and generally grungy appearance, and considered that the Rahrs might have cause for concern. But Stoddy, if anyone, understood that the path forward was not always straight. While the vet wandered aimlessly in the garden, Stoddy spoke plainly to Guido about the twists and turns of his own life and how he'd ended up as an academic. Guido liked Stoddard, who did not assert, as his parents had, that writing and reading were essential to a meaningful life, or insist that literature was the basis of all civilized culture, a premise that Guido flatly rejected. Stoddard made no such claims; he simply suggested that college was the best and easiest way to proceed.

After visiting the campus, Guido decided that the University of Oregon was looking better than street fighting and rough living in Bend—and there were lots of cute girls. There were also two fine trout rivers running right through town.

Nestled in the forested foothills of central Oregon, Eugene had been transformed from a sleepy lumber town into the "running capital of the world" by Olympic runner Steve Prefontaine. The city streets and rivers were lined with running trails populated by joggers wearing short shorts and the new and revolutionary Nike shoes. Guido pulled into town with a duffel bag of clothes, his fishing gear, two mountain king snakes, and a Burmese python. He had shoulder-length hair and wore his uniform of heavy woodsman boots, tinted glasses, and a wool hat pulled down low on his head. Walking the university halls on the first day of class, he noticed that he attracted more than a few stares. Next to the relatively clean-cut student body, he looked like a strange character from the backwoods. Inwardly he shrugged, and proceeded to find his own people at the local fly shop, the Caddis Fly, where he met some sufficiently lunatic fishermen who couldn't have cared less about how he looked.

Guido spent his first semester watching the student body the way he watched a patch of water—from a distance. He was unsure whether it was worth it to try to fit in, to apply his fly-tying talents to his own appearance and attempt to look like all the other bugs buzzing above the water. The Greek system of fraternities and sororities held no appeal for him, but on the edges of these monolithic social scenes were people who looked like they could be his friends. Guido decided he would at least pay the entry fee, and for the first time in his life, he physically adapted to his human environs. At a barbershop in town he had his long blond hair taken up above his ears. Then he wandered into the Gap and invested in another "uniform" that would last for years to come—brown jeans with a blue button-down shirt (three of each). The last stop was a shoe shop where he purchased white Tretorn tennis shoes. Cleaned up, Guido was fresh-faced and handsome. He could even have passed for preppy, the going look at the time.

With his updated uniform, Guido gained entry to social circles that would have barred him a week earlier. Social mingling was easier than he'd imagined. In fact, some buried memory of etiquette surfaced to smooth the way. He vaguely remembered his family's counsel that when attempting conversation, strive to be both interesting and interested. Guido followed the instruction and observed immediate results. He eased into regular socializing, finding that with a few beers in him and cute girls to look at, the painful exercise of chitchat was both uncomplicated and enjoyable. Part of him still felt like he was in disguise, but it was a disguise with benefits. At parties and pubs, the cute girls were taking notice of him. Besides the fact that he was physically attractive, there was something intriguing about him, a flicker of intensity beneath his easygoing manner.

Still, college failed to weaken his attachment to reptiles. In his sophomore year, he moved into the dorms, sneaking his mountain snakes and ten-foot Burmese python with him. That fall, he expanded his collection with a pair of young rattlesnakes he'd caught when visiting his pot-growing friends in southeastern Oregon. He secreted his specimens back to his dorm room, where they lived in aquariums with heated pads set underneath. The python subsisted on rats and frozen baby chickens, which he defrosted at the local 7-Eleven. The rattlesnakes consumed baby rats and mice, and the king snakes ate baby mice and lizards. Guido kept them all fed and sheltered, and all was well until the spring sun angled through the windows, stirring the snakes from their winter slumber with its warmth. Activity picked up as the king snakes explored the nooks and crevices of the dorm room, and the rattlesnakes rattled loudly from their terrarium. Guido took to playing loud music to muffle the sound.

One morning Guido woke up to hear a fellow student screaming, "GUIDO GUIDO! You bastard!" Guido ran into the hall to find a terrified boy clutching a towel around his waist. Coming down the hallway from the shower, he had caught sight of a long, red-banded snake slithering out from under his door—no doubt on its way home to Guido's room. But when the boy started to yell, the snake stopped. Sensing danger, it changed its mind, turned around, and retreated to the boy's room, where it disappeared from sight. Even Guido could not find it. In a frenzy, the boy emptied his room of every single object. Satisfied that the king snake was gone, he put everything back, and that night slept soundly. The next morning, Guido was awakened by his screams when the boy's bare foot landed on top of the harmless king snake, where it lay sleeping in a pile of clothes.

Guido found the whole episode sidesplittingly funny. When his hysterical dorm-mate calmed down, he comforted him with some old-fashioned snake wisdom: "Red touch yellow, kill a fellow; red touch black, good for Jack."

I remember being surprised when I heard that Guido decided to be an English major. Evidently his success at the High Desert Museum had given him an appreciation for storytelling and for the power of words. Cousin Stoddy had also endorsed the value of studying English. If nothing else, it would teach Guido to think. His parents were astonished to receive a chatty letter from Guido outlining his academic plans. "It was the first hopeful sign we had," Laurie remembered. His grades, however, did not reflect his burgeoning interest in great literature. Guido had other, more important things going on.

In Eugene there was a public-access television show called *On the Fly*—a twenty-eight-minute program about tying flies that involved nothing more than a table covered with clumps of feathers and fur. It was a quirky, popular show that aired weekly. Occasionally the producers featured guest hosts, and Guido's name was circulated as a possibility. The college kid was rumored to be an excellent angler—and, word had it, he could tie flies.

Guido instantly agreed; such a gig would lend much-needed relief to the monotony of school. For his guest spot, he decided to tie a wooly bugger, a staple in any fly box. There were many ways to tie a wooly bugger, for the robust black fly could imitate many things, but Guido had a certain style. Using marabou tail feathers, he opted for a bold pattern, with a shot of tailing extending out from the weighted black body. He would bind the fly with fine copper wire. Below the water's surface it would resemble a small minnow, or maybe a leech.

On the day of the shoot, he was ushered into a dark room with a fly-tying table, a chair, and a camera. The episode would be recorded in one take, so Guido had prepared to talk for the duration. He was exhilarated and somewhat daunted by the prospect of expounding for half an hour into a camera. Jumpy with nervous energy, he got himself situated at the tying table, where the little fly transported him far away. As soon as the camera started to roll, his nervousness dissipated and he was able to snap into the deep focused state he found on the river. Those watching recognized something electric in his presence. As Guido proceeded to carefully describe the attributes of the wooly bugger, his blue eyes flashed at the camera, transmitting not just energy and knowledge, but mastery. By the end of the twenty-eight minutes, the crew of *On the Fly* was silently respectful.

Guido did not have to be told that he had knocked the presentation out of the park. He was offered the job of permanent host, which he gladly accepted. School promptly fell a notch in his list of priorities, landing somewhere in the dust behind fishing, socializing, and, now, hosting a television show.

In the following weeks, viewership of *On the Fly* picked up. Even Guido's teachers started watching. He was occasionally called to stay after class not because of a lousy paper (though the lousy papers were there) but so that professors could grill him on the specifics of his angling techniques—had he used a size 16 or 18 long shank with that damselfly? Once again, Guido was able to sidestep the official curriculum and do exactly what he wanted to do.

With Guido at the helm, *On the Fly* gained a cult following and ran for decades. I watched them filming an episode once at the Deschutes. It couldn't have been less eventful. The set consisted of a small folding table and chair, on which Guido sat contentedly with a tying vise and his feathers and fur. It was all he needed.

Meanwhile, Guido was successfully ascending the fly-fishing ladder, unwittingly following the course of many fly fishermen, climbing from rainbow trout to steelhead to brown trout in search of the next most challenging fish. Recently a new rung had appeared on the ladder, and for many it was an impossible reach. Only a few fly fishermen were attempting to catch chinook salmon on a fly, and for obsessive anglers like Guido, it had become the ultimate fly-fishing test. Guido was determined to join the ranks of the few elusive fishermen who had passed this test, but he had no idea who they were, where they fished, or how. He had one friend who could induct him into their ranks, and he had been patiently cultivating this friendship for months, waiting for his chance.

There is a saying in some countries: "I'd rather give you my wife than tell you where I fish." So it was no small thing when, one October morning in his junior year, Guido got the call from his fishing pal Dave Witchey. Guido had done what he could to woo Dave, taking him to his secret pool of big browns, divulging his most coveted secrets. "Drop everything and get yourself to the Chetco River, now!" Dave instructed. "And get yourself a pram." Guido borrowed a pram (a small, flat-bottomed boat), packed his bags, and hit the road.

The Chetco River was four and a half hours away, due south. Guido drove through the autumn foliage with its blazing red maples, shimmering cottonwoods, and golden larches. At Crescent City, he turned north and hugged the craggy edge of the Pacific as he sped along Highway 101. The Chetco was a gorgeous river, flowing fifty-five miles from its headwaters in Siskiyou National Forest through rugged mountain canyons replete with towering redwood forests and waterfalls. Dave and Guido met up at its mouth and drove a couple of miles upriver, where they pushed their prams into the blue-green water. It wasn't long before they came to a calm, deep pool, where four prams just like theirs floated silently, anchors dropped on the river bottom below. The morning had the hush of church as Guido and Dave anchored their boats next to the senior anglers. When they were in position, Guido nodded respectfully at his esteemed companions and turned to survey the water surrounding them, which showed no signs of life. As he relaxed, though, his eye caught movement toward the bank. The water churned, and a large, dark back rolled to the surface. The prams were sitting on top of a huge school of chinook salmon.

The greatest of the Pacific salmon go by different names: king, spring, chinook. They are revered by those who hunt them for their strength and size, but mostly for how they fight. They are, simply put, the badasses of the salmon species. Honed by the ocean's deep currents, they swim the farthest and stay out at sea the longest, packing their bodies with muscle and fat before heading home to spawn. For the duration of this epic journey, they swim in fast, deep schools, rarely rising to the surface to feed—which is why they are so hard to catch. When they hit the freshwater of their birth rivers, they have one objective only, to spawn. As they make their way upriver from the sea, they are at the peak of their lives and the apex of their power.

The chinook of the Chetco are some of the biggest salmon south of Alaska. Starting in late September, they start to stack up off the mouth of the river, awaiting entry with the tide. Once in the river, the last and most arduous leg of the journey still remains. Some chinook have to travel as far as fifty miles inland to reach their birthplace, using the last of their strength to fight their way up rapids and leap up waterfalls. Before this strenuous final effort, the chinook of the Chetco find the gentle waters of the deep pools near the mouth of the river. And here, they pause.

Guido was awakening to the challenge before him. Light fly tackle,

small boats, and enormous fish promised drama. Dave informed him
that, once hooked, a decent-sized chinook could easily pull a pram
behind it, which meant you had to be a skilled boatman as well as an
expert fisherman to fight and land the fish without upsetting the other
anglers. Guido saw the delicacy of their arrangement in the pool. The
anglers would have to act in careful unison; one clumsily dunked oar
or upset boat would spook the fish for everyone. And while the fishermen were all casting together, they could only catch the chinook one
at a time. As soon as one of them got a fish on, the others would reel
in their lines while the lucky angler hooked, fought, and landed his
prize. This could happen one of two ways. If he could secure his rod
between his knees and manage to row his pram to the nearby shore,
the angler could fight the chinook from the bank. The more likely
scenario was that he pulled up his anchor and let the fish carry him
downstream until he reached water shallow enough to stand in.

Dave recommended a nymph, a steelhead fly that Guido knew
well. Nymphs sank far below the surface, where big spawning fish
like steelhead and chinook swam. Holding his place in the line of
prams, Guido slowly retrieved his fly in the beautiful, clear water,
casting his line toward the area along the bank where he had seen the
chinook roll. It didn't take long before the nymph was stopped by
something below. It wasn't a slow tug, like it would be with a big
brown. This felt like pulling a block of concrete. He wasn't sure it
was a fish at all, but he followed Dave's instruction to set the hook
hard. A moment later, the rod was nearly ripped out of Guido's hands.

The chinook was the strongest fish Guido had ever fought—and it
fought like hell. It took every muscle in his body to hold the rod
steady as the fish dove, flashing in the deep water as it ran, again and
again. It was Guido's first encounter with a fish that would simply not
give up. He could feel the urgency, the pure life force at the end of his
line. Guido won the fight in the end, but it felt more like a stroke of
luck than a victory. Staring at the huge silvery fish at the bottom of
his pram, he was filled with respect. The chinook was beautiful, a
gleaming messenger from another world. Who knows where it had

been, and what it had seen. Guido sat back, done for the day. His goalpost had moved, and it would not move back.

Chinook salmon quickly surpassed the status of the browns of the upper Deschutes, for here was a fish that demanded everything of him. Guido could not begin to imagine what he would have to learn to be equal to such a creature. Its habitat wasn't just a pool in a river; it was the entire Pacific Rim. These fish had traveled thousands of miles, feeding in distant, cold seas. Some, Guido had heard, swam as far as Russia. Understanding chinook would require deep study, perhaps years of it. It would also require a pram, two oars, and a trailer.

At the end of his senior year, Guido decided his plans to chase chinook had to wait. While he had discovered the joys of literature, some things hadn't changed. His lax participation had landed him in the same spot he'd found himself at the end of high school—without a degree. He would need to attend to a neglected language credit in summer school. Guido off-loaded his snakes to a postdoc in the neuroscience department, keeping only the one he could not part with, a reticulated python named Monty. After being kicked out of the dorm, Guido and Monty lived as vagrants while he spent a few hours a day in Spanish class. Guido already spoke the language, so the class didn't require much attention. In fact, the final presentation slipped his mind completely. Guido was up the night before partying with some friends when he remembered it. Having no idea what to do, he looked down at his gym bag, where Monty slept peacefully. The one thing Guido could talk about spontaneously was his reticulated python.

The next day, he slung the gym bag over his shoulder and headed for class. It was the dog days of summer and most of the students in the warm room were half asleep. Guido was the last to present. He stood up in front of the class with his gym bag and said, *"Mi nombre es Guido, voy a presentar mi serpiente."* A few of his classmates stirred, recognizing the word *serpiente.* Guido explained that "retics" were technically the longest snakes in the world. Monty was a teenage retic, about thirteen feet long, with a head the size of a Brittany spaniel. She was a big, good-natured snake that stayed happily curled up in the gym bag for most of the day. Guido fed her by dropping a rat into the bag. The bag would shake once and the rat would be gone.

In the classroom, Guido could feel that Monty was getting active. When he finally unzipped the gym bag to display the object of his presentation, the snake shot out like a geyser and headed straight for the light fixtures on the ceiling. Guido continued speaking as he calmly tried to pull her back, coil by coil. Monty wound herself around his shoulders as he described her feeding patterns, rearing up above him to survey the room. By now the entire class was flattened against the back wall, inching toward the door. As Guido attempted to assure their safety, his classmates escaped into the hall, leaving him and Monty alone in the abandoned classroom.

After squeaking through Spanish, Guido packed his van and drove off, towing his new white pram behind him. As far as he knew, the attempt to catch chinook on a fly was being carried out only on southern rivers, so Guido headed north. Up north they didn't know about prams, or trying for chinook with a fly. He would be alone in virgin territory. For Guido, it was perfect.

Three hundred miles north of the Chetco was the Tillamook Basin, where five rivers ran fast into the Pacific. In the town of Tillamook, Guido stopped to make inquiries at the first fishing shop he saw. As he had thought, chinook passed through these waters, and there were no fly fishermen trying to catch them. The question itself drew laughter. Chinook swam too deep and fast, and were too strong for fly-fishing tackle. Besides, they rarely fed on anything, much less a fly. Guido knew he had found the right spot.

In the next weeks, he befriended a group of gear fishermen who trolled the mouths of these coastal rivers for the king of salmon. Some of the fishermen were also guides and knew the rivers well. They also knew where the fish were—and they weren't easy to find. Guido learned that 5 percent of the river held 95 percent of the fish. You could spend all day fishing and not catch anything if you didn't know where to look. And there were other factors to consider: when the water level rose or fell, a frequent event on a coastal river, the fish might move to a more advantageous spot. There was nothing fixed about this ecosystem, it was constantly in flux. This was all useful information that Guido filed away for a later time, when he would need it. For now, he remained an uncomfortable apprentice.

Guido didn't like the techniques of the gear fishermen. They used bait, which he didn't approve of, or a "hot shot," an artificial silver minnow with a flat bill that wiggled and dove from where it was tied thirty feet behind the boat. When the fishermen rowed their drift boats against the current, the hot shots swam and dove underneath like crazy drunken pilots. The boats rowed back and forth across the pool, agitating every patch of water with their underwater irritants. The hot shots also had rattles inside them, and generated a noise that annoyed the fish to no end. Guido noticed that chinook had a temper. He watched gear rods go from standing upright to flat when a chinook grabbed the lure. The fish, no doubt fed up, struck hard at the cause of its irritation. Then, as the two huge sets of treble hooks dug deep into its mouth, the fish ran like hell, the line not so much whirring as screaming away. To Guido, it seemed an undignified way to engage with such a majestic creature.

He kept his opinions to himself, however, as he was here to learn. On the gear boats, he was at least able to get close to the chinook, and he never tired of observing them. The gear fishermen let him tag along, tolerating his endless questions and incessant curiosity because he was good-natured and could put up with the perennial discomforts borne by serious fishermen, which, for Guido, included sleeping on the floor in the filthy basement of the bait shop. He was on the hunt, and it was going to be a long one. When he was ready, he would leave the gear fishermen and find his own water. He would fight chinook fairly, giving them odds with a small fly and a small boat, meeting the mightiest of salmon on equal ground.

It was in the fall that Guido left the bait fishermen, venturing upriver to explore areas that many had given up for dead. The rivers that ran through Oregon's steep coastal range flowed through a temperate rain forest that once held the oldest stands of Douglas firs on earth. It was a spectacular forest with towering 1,000-year-old trees that, fifty years earlier, had burned to the ground.

Like all Oregonians, Guido had heard tales of the unholy conflagration. Our parents, children at the time, had watched the glow of the Tillamook fire from forty miles away in Portland. Oregon's forests had been heavily logged for decades, but in the month of August 1933, the state had issued a warning for lumber companies to suspend their operations, for temperatures were high and humidity levels were dangerously low. But in a remote canyon, one timber company ignored the warning, dragging a last Douglas fir across the forest floor with its mechanized winch and cable. When the gigantic tree rubbed against a desiccated snag, the friction produced a small, white plume of smoke.

It wasn't just any fire. Some believed it was the forest's revenge for the aggressive and careless cutting of these ancient trees, for little else explained the ferocity of the burn. Douglas firs were some of the most fire resistant of the big evergreens, and were generally scorched only by nature's cyclical burns. But the Tillamook fire reduced these magnificent giants to ash. Photographs show something that looks more like an atomic blast, with thick stacks of mushrooming clouds lit from within by an unearthly gleam. For the weeks the fire raged, coastal communities lived in darkness as singed needles, branches, and cinders carpeted their streets. In the town of Tillamook, the air was so thick with ash that chickens roosted at noon. Far out in the Pacific, ships reported falling debris from five hundred miles offshore. Aside from its incendiary heat and scope, the fire had an eeriness, and there were reports of things people couldn't explain. Aerial accounts described bomb-like explosions and old-growth trees being uprooted and hurled great distances.

By the end of August, the Tillamook Forest was a desert of smoking embers. The birds and wildlife were gone from the trees, the fish gone from the rivers. For Oregon, it was a travesty; they had lost 300,000 acres of one of their greatest natural treasures. Lumber companies walked away from the disaster, leaving the mess for the state to manage, though there was nothing much to manage; there was nothing to do but start over. Our parents, along with many local schoolchildren, were part of the restoration effort, trundling out in school buses to dig into the burned earth and plant bright-green baby Douglas firs.

It is said that trees in the Tillamook Forest grow faster than any trees on earth. Every year, more than one hundred inches of rain run down its lush, steep hills, filling gullies, cracks, and crevices that feed rivulets, streams, and creeks that move through the land like a circulatory system. Over the next years, the little seedlings planted by our parents took root and grew, and grew. When Guido discovered it fifty years later, the forest was thick with new growth. The baby Douglas firs now pierced the sky at 130 feet.

He stood alone on the banks of the Kilchis River, noting that interspersed among the evergreens were red alder, maple, and hemlock. The air was fresh and cool from the ocean, and he breathed in the fine, salty mist as he watched for signs of fish. Then he pushed his pram into the crystalline water and bobbed along for a minute or two. The pebbles below were perfectly clear, the sky above cobalt blue. He felt like he was floating on air. As his eyes relaxed on the water, every now and then he saw a shape moving beneath him, some as long as his arm. By the end of the day, he knew the forest had been reborn—and with it had come the fish.

For the next year, Guido haunted the rivers of the Tillamook Basin, and he haunted them virtually alone. He was astonished by what he saw. Every river was filled with salmon—each with its own array of species. On one river, he observed four races of chinook salmon that surged in from the sea, one after another, starting in spring and ending in winter. Other rivers had steelhead, coho, and chum salmon, which surprised Guido, for these fish needed extremely clean water, and were fast disappearing from the Pacific Northwest. Guido had fished many world-class rivers, but the watershed in his own backyard was equally fine. And no one seemed to know it.

As he watched the fish, tracking their progress to their spawning grounds, Guido saw there was nothing random to the order of their arrival time. It was like a choreographed dance. The salmon that came in first had the farthest or most difficult distance to cover, with obstacles like waterfalls to overcome. These were the spring chinook that spawned all the way at the top of the river, at the headwaters. Because it took them longer to reach their spawning grounds, they entered the river sooner. Depending on the river, coho or summer chinook came next, spawning just below the spring chinook. Then came chum and sometimes a fall run of chinook, which might be followed by steelhead, or a winter run of chinook, each finding its slot to fill in the lower sections of the river. Guido realized it was a ladder of fish, starting from the headwaters and moving systematically down the watershed.

He was struck by the elegant economy of the species, how each salmon grew exactly as big and strong as it needed to be to reach its home waters. On the Nehalem River, chinook were at the top of the ladder. These were the fish that swam the farthest and had packed on the most muscle and fat to fuel their arduous journey. Steelhead were right behind, with sockeye and coho following. The salmon that traveled shorter distances, like pink and chum, had soft flesh with little fat to fuel them and a meat that began to disintegrate almost as soon as they hit the freshwater. Spawning just inside the river's mouth, they didn't have far to go.

At spawning time, each species changed in its own particular way, turning bright red, brown, or green, or gaining stripes. The males of some species developed a hooked nose and a humped back or sharp teeth, changes that helped them attract a mate and defend their spawning ground. It was almost impossible to mistake the species at this stage, as they were totally distinct, and while they were all swimming upriver, they were headed to different places.

Guido regularly stalked the banks of six rivers, watching as salmon species swam in one after another, filling the river from top to bottom. The closer he looked, the more miraculous the salmon ecosystem became. It seemed each species was programmed to reach a specific spawning ground at a specific time of year. The logistics of this were staggering. Somehow salmon coordinated their sexual maturity with varying water levels, temperatures, traveling time, and difficulty. It was a complex equation that, when it worked, resulted in the successful bearing of offspring. Later, genetic analysis would show how precisely these fish had adapted to their home waters. For example, those that spawned in rivers with waterfalls had expressed a distinct "jumping" gene. But even now, Guido could see how every fish fit snugly into its place in the ecosystem. It was a beautiful jigsaw that had been tens of thousands of years in the making.

That the Tillamook Basin was a salmon stronghold came to Guido gradually, piece by piece, and it came with his understanding of its rivers. Unlike many rivers to the north and south, there were almost no dams here, and most of the fish were still wild. These were healthy, unimpeded rivers, and salmon were able to utilize every piece of their habitat. While there were threats from lumber companies and dairies, these threats were not yet fatal. But salmon would always face threats.

The beauty and the tragedy of salmon, Guido thought, was that their life journeys forced them into human hands at the most critical moment of their lives. From the moment they entered their home rivers, salmon faced obstacles, most posed by humans, and some impossible to overcome. The story of Atlantic salmon sat uneasily within him. It was not so long ago that salmon had filled the Rhine, the Seine, the Thames, and rivers from Portugal all the way to Norway. These fish had been a mainstay for Atlantic peoples since ancient times, and were reverently depicted in cave paintings. Monarchs and governments had protected salmon for centuries, instinctively understanding their value. There were other illustrious champions—Charles Dickens wrote in 1861, "The cry of 'Salmon in Danger!' is now resounding throughout the length and breadth of the land. A few years, a little more over-population, and a few more tons of factory poisons, a few fresh poaching devices . . . and the salmon will be gone—he will be extinct."

Dickens's prediction came to pass. It had taken one hundred years of aggressive overfishing, pollution, blocked rivers, and logging for European salmon runs to dry up—one hundred years of a war on all fronts to defeat them. While the Pacific coast was far from such a demise, Guido was recognizing that what had happened to Atlantic salmon could happen in the rivers he was standing in now. Salmon were strong and resilient, but denied a few basic needs, they couldn't survive. The fact was, they didn't need that much. They needed clean, cold water and a way to reach their spawning grounds. But the Pacific Northwest was filling its salmon rivers with dams, causing agricultural pollution, and clear-cutting the forests.

Guido had heard that the lumber companies would soon return to Tillamook, for these young trees were almost ready to harvest. As he stood in the misty morning waters, with the dripping of Sitka spruce and the smell of the sea, he protested inwardly at the prospect of this new forest being razed. It wasn't just the violent disruption or the ugliness of the clear-cutting scar; he had seen firsthand how reliant salmon were on trees. Their branches offered cover, and their roots and branches provided structure: fallen logs created breaks in the current and places for baby fish to shelter. From saplings to toppled giants, trees were the backbone of salmon rivers. Removing them was like pulling the A-beams out of a house—the supports without which the house would collapse. When the lumber companies started clear-cutting again, this miraculous little ecosystem would be in trouble.

Guido spent as much time as he could in this secret watershed, trying to learn everything he could while he had the chance. He felt fully alive here, following the salmon as they came and went, gradually coming to know when they traveled where. It was a close and patient study that resulted in a deep understanding of a phenomenally productive ecosystem. For Guido, this was a gift like no other. It was also a weapon that he would learn to wield.

CHAPTER 4

THE CITY

FLY-FISHING WAS NOT A JOB, but Guido tried hard to make it one. He worked briefly as an outdoor reporter for a television news station, then made fishing videos with a friend. Neither venture amounted to much. It was his cousin Spencer Beebe who saw that while Guido was uncovering the secrets of his local rivers, he was floundering professionally, and that his talents could be better utilized elsewhere. He suggested that if Guido loved the natural world, he should learn something about protecting it.

Within the Nature Conservancy, Spencer had developed an international program protecting biological hot spots, zones that represented earth's healthiest ecosystems. While not necessarily large, these areas possessed both an unusual abundance and diversity of life. For scientists, the intricate balance and intelligent order of such unbroken ecosystems were invaluable; the study of them had the potential to illuminate much about the interconnected nature of life itself. Intuiting their value, Spencer was one of the few Americans in the race to protect these extraordinary places before they vanished.

Guido innately recognized the value of any wild place, and he was happy to fight for it, though this rookie assignment would test him. Spencer dispatched Guido far from his beloved northern rivers deep into Mexico, to the cloud forest of the Sierra Madre. Near the remote region of Chiapas, the rain forest was burning, spreading its flames toward lone swaths of mountainous land. In the clouds of these mountains was a forest ablaze with orchids, bromeliads, and giant butterflies. The branches of its ancient oaks harbored four hundred species of birds, among them the horned guan, the azure-rumped tanager, and the mythical quetzal. The cloud forest was a beautiful floating island, surrounded by smoke and fire. The locals were steadily burning the lush habitat to create farmland; 90 percent of it had already been lost.

It was a broiling and beleaguered part of the world, but Spencer believed that Guido was capable of handling such a stint. He thought Chiapas, with its plethora of snakes and reptiles, would suit his cousin fine. It was a hardship post—both uncomfortable and dangerous. As tension built between the natives and the landowners over the natives' right to farm the forest, subcomandante Marcos led his Zapatista army deep into rural regions to defend the native people who needed to farm the jungle to survive. Traversing the mountainous region on horseback, he had organized an army of Mayan farmers to resist the armed *federales* who had been sent to stop them. Marcos's agenda pitted him against environmentalists, who he suspected were in collusion with corporate interests, if not the government. Guido was to expand the skeleton program the Nature Conservancy had in place. There were no gringos in the region, Spencer told him; Guido would be on his own, without support. These details faded into the background for Guido. To him, Chiapas translated into "reptile central."

In 1985 he flew down to Tuxtla Gutiérrez in his seersucker blazer. When he got out of the airplane his new boss, Ramon Perez Gil, was there to meet him. Gil took one look at Guido and said, "Iyiyiyi, you're not going to need that coat." Guido was promptly swallowed by 113-degree heat and air smoky from the burning rain forest.

In Chiapas, there was no possibility of blending into his

surroundings—blond-haired, blue-eyed Guido stood out like a sore thumb. Children cried when they saw him. He was used to being an outsider, but in Chiapas his isolation was profound. He was stationed at the Zoológico Miguél Álvarez del Toro, a 250-acre reserve that harbored some of the last remaining species of regional fauna and wildlife, and his job involved expanding the protected areas of western Chiapas and increasing awareness about this unparalleled habitat. A one-man team, he was also tasked with raising money and building support for Mexican conservation groups, which required cooperating with and hiring locals. Fluent in Spanish, Guido was keen to understand their situation, certain they could find common ground. The locals, however, did not trust Guido. Nor did they really understand what he was doing there. They could see he was crazy for reptiles, but chasing lizards could not have been his sole purpose in coming to Chiapas. What grown man did this?

More plausible was that Guido was an agent gathering intelligence on the unrest in Chiapas. Mexico was on the verge of a civil war, and Guido could have been working for either side, though some thought he was employed by the CIA. Complicating matters was that soon after Guido arrived, the Nature Conservancy underwent a dramatic split in leadership that culminated in Spencer Beebe and his colleague Peter Seligmann leaving and forming a new organization called Conservation International. CI was Guido's new employer. Wasn't CI a thinly veiled reference to the CIA? No one in Chiapas had heard of Conservation International, but they had all heard of the CIA.

Guido chose to remain ignorant of the conflict that surrounded him, focusing instead on hunting reptiles. This tropical zone yielded a plethora of intriguing snakes, many of them poisonous, and he was eager to catch as many as possible. He used a two-stick method whereby he immobilized the snake's body with one and its head with the other. Once he had stabilized the head, he could pick the snake up and take a closer look. He had a few close calls, one with a coral snake whose tail looked so much like its head that Guido didn't know he had the wrong end until the snake had sunk its teeth into Guido's stick.

Day by day Guido learned the business of conservation. He took photographs and made brochures to distribute when he gave fundraising talks in the United States, where, it turned out, no one even knew the Mexican cloud forest existed. In Tuxtla, he boldly reached out to powerful individuals, the big landowners, and found he knew how to talk to them. He introduced them to Ramon Perez Gil and started building bridges that connected the locals to the larger world. Gradually, the locals began to see that Guido had no secret agenda and was genuinely interested in understanding their plight and finding solutions. Alejandro and Alexander, the Mexican biologists at the institute, were entertained by Guido, and came to like him.

The unrelenting heat was brutal on the boy from North America. The days were 110 degrees and higher, and Guido passed from one to the next as if in a fever dream. Physically, he had never been weaker. Parasites plagued his system, curling him into a fetal position with agonizing cramps, until finally he took himself to the local pharmacy, where he got a shot in the bum that temporarily alleviated his symptoms. Later he regretted not being more careful about ingesting water, though avoiding it altogether was impossible—it was in the ice cubes of his drinks, in the sink when he brushed his teeth. After six months, his immune system began to suffer, leaving him vulnerable to more serious illnesses.

It was May when he and his two Mexican colleagues were sent to explore the beautiful, remote rain forest of El Ecote, to take stock of the forest and to see how far into the park the locals had cleared it for their corn and bean fields. They also wanted to find a certain cave, rumored to have Mayan artifacts. As the three men headed into the interior of the park, Guido felt the beginnings of a fever. As they continued deeper into the forest, with its ancient trees and thick mossy ground, Guido started burning up. He was sick, and getting sicker. At one point the forest started to spin around him, and the world went blurry. The next thing he knew his Mexican colleagues were half carrying, half dragging him along the trail. Guido insisted that he was fine and that the others should keep trying to look for the cave—at which point his colleagues deposited him in the shade of an oak tree, covered him with leaves, and set off for the cave. Guido lay curled up under the leaves, shaking and too sick even to scan the canopy for harpy eagles, his favorite raptor. He drifted in and out of consciousness as the hours passed. It was evening when his friends returned, uncovered him from the leaves, and hoisted him to his feet. Propping him up between them, they shuffled him back to the car. When they got back to Tuxtla, Guido collapsed into bed with what turned out to be a virulent case of typhoid fever.

Guido spent his negligible time off healing his body and soul in the rivers of the Pacific Northwest. He was widening the radius in his search for big fish, which were increasingly hard to find in his home rivers in Oregon. Flying north from Mexico, he had seen the reason why. In the land below was a pattern of disruption, starting with California's great Sacramento River. Once as mighty as the Columbia, the Sacramento had been diverted, leveed, polluted, dammed, and altered beyond recognition. In the vast Central Valley now were brown, barren tracts and fields upon fields of fruit and nut trees irrigated by redirected river water. The richness of life on the river—the birds, fish, amphibians, and vegetation—had been winnowed down to a few lone survivors. The chinook that had filled its waters were all but gone. In Oregon, Guido realized that something similar was afoot. Land that had once been defined by forests and watersheds was now chopped into many little squares, some belonging to the state, others to lumber companies, others to farmers. This patchwork of dysfunction could spell disaster for the salmon. Denied cold, clean water and a path to their spawning grounds, they were unable to reproduce—and often died trying.

There were still fish in the north, plenty of them. Guido had followed them from Washington State to British Columbia, all the way up to Alaska. The farther north he went, the bigger the fish. Every chance he had, he headed to Alaska's clear glacial rivers that teemed with salmon—big, beautiful fish strong from their journey across the Bering Sea.

It was a few years later that Guido and I both found ourselves living in Washington, DC. I was working at *The Washington Post* and in my free time studying Russian. After three years in the jungle, Guido had been promoted to Conservation International's headquarters, where he was unhappily bound to a cubicle. Half-blinded by concrete and glass, he spent his days imprisoned in an office on the fourth floor with sealed windows. He knew at once he was in trouble.

Spencer Beebe and Peter Seligmann were also trapped in these offices, though they didn't seem to mind. Guido studied them, noting with some dismay that his mentors were clad in coats and ties and spent much of their days on the phone. While Guido had proven his ability to build a conservation program in the field, Spencer told him, there was another side to the business. If Guido wanted to continue at CI, he needed to raise enough money for the cloud forest to secure its protection. If he failed at this, they would have to let him go. They weren't talking about a few thousand dollars, Spencer clarified; they were talking hundreds of thousands.

Both Spencer and Peter had doubts that he could hack the assignment, and Guido knew it. As of yet, he hadn't displayed terrific people skills. It was a test, perhaps the hardest one yet, for there was little to sustain him here. Instead of working with nature, he'd be working with people, all day, every day. And the earth was covered with concrete.

But the challenge activated something in him. Guido was a fighter, and he had adapted before. He bought the suits and ties, deciding against the dress shoes, which were surely a waste of money. He found a little apartment in the Adams Morgan neighborhood of DC, rose early every day, and walked to work, arriving before anyone else because a colleague had informed him that to succeed at CI, he would have to be the first one in the office and the last to leave, advice he took literally. It made for excruciatingly long days in an environment he found almost unbearable. To remain sane, he took up running, hitting the streets sometimes as late as ten P.M., just to feel his body pulsing through space and purge his lungs of stale office air.

In his cubicle, he hung a miniature terrarium, a ball of orchids and moss that mimicked the cloud forest. Invisible to everyone but Guido were two lizards that lived on the cluster of greenery. They were Bahamian anoles, a small green lizard with sticky feet and the chameleon-like ability to change color. He stared at this tiny orb of wilderness and watched the anoles silently explore their new world, envying their ability to fit in.

While Guido somewhat looked the part, his new role eluded him. The nuanced dance of fundraising and the people of Washington also eluded him. In this highly socialized realm, actions and motives could be strategic to the point of obscurity. Guido was straightforward, if not transparent. Charming potential donors over dinner at fancy restaurants was beyond him. He was confounded by office politics, and had no idea how to be on a team. His coworkers learned that he was physically incapable of sitting through planning sessions, and he got up to pace or came up with reasons to excuse himself. The real problem was that it had become his job to appeal to the most complex organisms on earth, and he had spent little to no time studying them.

Guido took his cousin Spencer's advice, put his head down, and focused on what he did best. He assembled a slide show with photographs of the cloud forest, piping in rare recordings of howler monkeys and sounds of the jungle. Then he took this presentation to clubs and organizations across the Midwest and Pacific Northwest, making the best case he could for the protection of such places.

It wasn't a total failure; Guido could feel he was connecting. But while his audiences were often stirred, the cloud forest of Mexico mattered little to them. They had never heard of Chiapas and had no plans to go. For these people, Guido's slide show was an interesting way to spend an evening, but they were not inspired to write big checks. At best, he got checks for a few hundred dollars—enough to cover his plane fare.

Back in the office, Guido faced the doubt of his coworkers, who gave him wide berth as he visibly struggled to sit through meetings and attend to organizational minutiae. It seemed the only thing they saw in Guido was his failure to bring in money. For his part, Guido tried not to be annoyed by their general ignorance about the earth they were trying to protect. Many of them would not have survived a day in the field, and had little idea of what they were talking about when it came to species or habitat. Yet Guido knew their skepticism at his ability to survive in this ecosystem was warranted. Six months into his new employment, his boss told him the jury was still out and described Guido as a "caged animal."

Guido stared at his anole lizards absently, comforted by the small bit of nature. Looking more closely, he noticed the anoles had bred. There were now baby lizards darting around their tiny open-air terrarium. It was a week later that Guido noticed a squashed baby anole near the entrance of the office, far from its terrarium. Somehow the little creature had made it all the way across the office, no doubt trying to find its way to fresh air. Yards from the door, it had been flattened by a shoe. The lizard's death saddened Guido for days.

Socially, though, he was having better luck. Guido had struck up friendships with employees of the World Wildlife Fund and National Geographic—people who had at least heard of cloud forests, salmon, and Alaskan rivers. Guido impressed them with his knowledge, which easily eclipsed their own. In a city of suits and graduate degrees, Guido was a curiosity, calling his friends to gather when he returned from one of his frequent fishing trips to Alaska or the Pacific Northwest with an ice-packed chinook. They feasted on salmon while listening to his tales of the catch, and descriptions of distant lands. While many could not articulate why, they were drawn to him. Women especially.

Guido was grasping that he was attractive to women, but he had yet to learn their ways. He did not initially see the conflict in spending time with two girlfriends—especially if one lived in New York and one in DC. He was naïve enough to think that these women might be open to hearing about the existence of the other, but it wasn't okay—with either of them. This baffled Guido for some time, but ultimately he accepted the protocols of romance and stuck to one girlfriend at a time.

For all his social gains, Guido felt like he was living on borrowed time. His trips to western rivers only temporarily replenished him.

During the silences that often filled his days, he began to listen more closely to the goings-on next door in Peter Seligmann's office, a seemingly magical, moneymaking place. Peter understood people and social systems like no one Guido had ever known. He noted Peter's affable tone as he roused potential donors to the beautifully urgent cause of conservation, inspiring them to join a fight that would spring them from the banality of their lives and allow them to become part of something greater. Once their maps had been redrawn and expanded to a planetary level, once they saw the part they could play in the most important drama of their time—he had them. The lines he spoke weren't in the least bit scripted; he was fully and completely present with his message, which felt original with every iteration. His conviction was totally genuine, and it was irresistible.

Guido recognized this particular charisma, for he had felt it in himself at the High Desert Museum and in front of the camera of *On the Fly*. He knew he was capable of channeling that bright, pure energy, of merging with his story as he described what mattered most to him. But he couldn't achieve this state of transcendence in an office with sealed windows; he couldn't do it on the phone. Still, certain truths were presenting themselves, and Guido could feel he was on the verge of a breakthrough.

Around this time we met up one night for a drink in Adams Morgan. I barely recognized him with short hair, khakis, and white button-down shirt. He looked put together, even dashing. When I raised my eyebrows, he marched to the bar and ordered two shots of mezcal. When he sat down I saw that he was still himself, barely containing a knee-jouncing restlessness as his eyes roamed the bar, searching for movement that wasn't there.

He laughed as he told me that the only fish he found were in the reflecting pools of the Washington Monument, where he'd discovered a healthy population of largemouth bass. He fished for them regularly, occasionally hooking a tourist on the backcast. Once or twice ducks had taken his lure and flown off with his line.

We had another shot of mezcal. As he talked, I became aware of other transformations. For decades, Guido had been in the habit of obscuring his eyes behind hair, hats, and tinted glasses—or all three. One could barely see his quick, appraising gaze, or feel the insult of its brevity if you failed to interest him. But he held my eye steadily now, and there seemed to be a new confidence in him.

Finally something had clicked for Guido, a pearl of wisdom he'd gleaned from Peter Seligmann. Success in this field was dependent on how you felt inside. If you weren't feeling the beauty and vulnerability of the cloud forest, you couldn't project that to others. If you weren't feeling the mission in your heart or you were just having a bad day—you might as well not bother, because people could sense these things. To inspire them, you had to feel inspired. Wait for the feeling, Peter advised, or learn to create it.

Guido realized that instead of thrashing it out in a system he had no feel for, he had to find a way to bring people into *his* world. If he could transport potential donors to a beautiful, intact ecosystem, he was sure he had a chance at making them care about it—maybe enough to write a check toward its protection.

Guido waved down a waiter and ordered from the Tex-Mex menu in Spanish. He smiled as he told me that he'd tried this new strategy on a trip guiding potential donors to the cloud forest. It had been a gamble that had almost ended his career, but it had also taught him something critical about human nature, a lesson that would become the cornerstone of one of the most radical and successful fundraising strategies in the conservation world, and one that was unique to Guido.

Out of desperation, he had turned to the connections he'd had growing up, the wealthy families of Minnesota. This was the country-club set, the insiders who had once shut him out. None of that mattered now, for these people were appearing to Guido in a new light, not as symbols of the world that had rejected him, but as potential allies. He decided to go for the brass ring straightaway. His target was an old family friend, Sam Bell, of the General Mills fortune. From a cellphone at the horse races in Vero Beach, Florida, Sam responded to Guido's heartfelt offer to show Sam and his wife, Susie, natural beauty like they'd never seen before. Guido described his unusual access to the extraordinary and endangered habitat of the Mexican cloud forest. It was a rare opportunity, one that wouldn't last. Guido assured them it would be an easy trip, and urged Sam and Susie to bring anyone they liked—it was only a short walk to the forest where they would see giant butterflies and resplendent quetzals. Sam and Susie were happy to make a party out of it, and invited family and friends from their well-heeled horsey world of Vero Beach. Guido was delighted with the numbers. He couldn't see what could possibly go wrong with taking nine socialites to the cloud forest of the Sierra Madre.

A little like the dog who caught the car, Guido hightailed it to Chiapas to make preparations for a group that would no doubt need special treatment. It was then that his Mexican colleagues reminded him that it was an arduous, three-day march to get up into the cloud forest, a detail Guido had neglected to divulge to the Bells. The most Guido could do was get the horse people some horses. They weren't the best horses on earth, but they would carry his guests up the mountain.

A few weeks later, the Bell party arrived in Tuxtla, where nine of them in varying degrees of physical fitness disembarked from their plane and were hit with 110 smoky degrees. Guido noticed they were decked out in polo shirts and sun hats. He watched uneasily as a parade of designer luggage and duffel bags clanking with bottles of alcohol was unloaded from the cargo hold. The Bell group seemed a little stunned by the smoldering landscape. Susie Bell approached Guido. "Where's the cloud forest?" she asked politely. "Where are the resplendent quetzals?"

Guido gently broke the news. "Well," he said, "we've got a little walk ahead of us." Then he smiled and told them how glad he was that they had come. In fact, after a five-hour drive to Tapachula, it was seven thousand feet to the top of the mountain range.

Nothing was as Guido had said it would be. There were so many bags that the horses (the Bells called them burros) had to carry the luggage while the Bell group walked. But Sam and Susie and their friends had come to have a good time, and the group was decidedly cheerful as it set out, even after an unsettling breakfast of shrimp (that had come from god knows where) wrapped in some kind of wilted leaf. Scurrying alongside them was a gaggle of scrawny chickens, which, Sam guessed, would serve as food for the trip. The going wasn't terribly tough and the party chatted and joked as they labored in the heat, fortifying themselves with beer and sips of scotch along the way.

The first night, they camped in the foothills, in the scrub forest. The Mexicans set up tents and sleeping pads while Guido lit a big fire. He was arranging logs so the group could sit, when he eyed a suspicious-looking piece of bark on one of the logs. Lifting it up, he exposed a five-inch poisonous scorpion, which he calmly escorted into the forest. He considered more deeply the inhospitable nature of the environs. The forest was full of scorpions and tarantulas, not to mention countless species of poisonous snakes. These were snakes he hunted; it was hard to think of them as dangerous. But dangerous they were. There was the coral snake with its paralyzing neurotoxic venom; the aggressive bushmaster; the irritable fer-de-lance, whose hemotoxic venom caused profuse internal bleeding; the jumping viper; and eyelash pit vipers that hung from prehensile tails and favored low-hanging branches. There were many ways to die on this lighthearted expedition, it dawned on Guido. At this point there was little he could do about it except keep the Bells close to camp.

The Bell party was happily oblivious to the dangers that lurked nearby. The cocktails flew around as the Mexicans looked on with no small amount of worry. Guido left the merry group by the fire and retired early, hoping they would all live till morning.

Instead of going to bed, though, the Bell party decided to explore. Half in a dream state, Guido thought he could hear them wandering around in the snake-infested scrub forest and drifted back to sleep. The Bell party walked far enough to find a river where, emboldened by the moonlight, the women decided to take a swim. They stripped their clothes off and waded into the water, laughing and splashing around, cleansing themselves of sweat and dust.

The Mexican guides had not gone to bed and were watching the progression of the group nervously. When the crazy gringos were all naked in the water, Alejandro hightailed it back to the camp and yanked open the zipper to Guido's tent. "Guido, Guido," he whispered. "Your guests are skinny-dipping in the river. They can't be wandering around like this—it's too dangerous." Guido roused himself. His friend was right. In addition to harboring snakes and scorpions, the region was semi-lawless, right up against the Guatemalan border. Guido got himself up, they gathered some towels, and together they herded the drunken, giggling guests into their tents and put them to bed.

The next morning, the Bell party was alive, but physically and mentally impaired. Staring at the chickens, Sam saw that their numbers had been reduced by half. Had they eaten chicken last night? He didn't think so. Badly hungover, the group listened queasily as Alejandro admonished them for their late-night frolic. Many dangerous creatures came to life in darkness; this was not a place you could roam around freely at night.

Chastened, the group ate what breakfast they could stomach and started off on what might have been a trail. Really they were bushwhacking. When they reached the base of the mountains, the terrain quickly tilted skyward. Soon it seemed like they were climbing straight up, navigating small cliffs and clambering over boulders. And it was hot as blazes.

Guido noticed that his guests weren't managing very well. Some of them had diarrhea and others were stopping to vomit in the bushes. Sweat poured down the face of an out-of-shape lawyer (whom I will call Marty), who was lagging behind. Guido bounced up and down the line offering encouragement. His boundless enthusiasm for the habitat buoyed Sam and Susie, who, grateful for the distraction, listened to Guido explain how, dispersed by birds, the seeds of a strangler fig germinated in the nooks and crannies of other trees, and from there grew both down to the ground and up toward the light. Guido pointed out toucans perched in trees, with their stout bodies and bright, clownish beaks. "Wait till we get to the top," he told them. "You won't believe it."

Alejandro kept the pace up; they were trying to reach a waterfall that would serve as a beautiful campsite, with soft, level ground. But they were already behind schedule, and as they pushed forward, coaxing the Bells as they went, no one noticed that Marty had disappeared from the back of the line.

As darkness fell, they were still a mile or more short of the waterfall. They would have to strike camp on a scrubby hill and manage the best they could. The Mexicans cleared the vegetation with their machetes, hacking away at the small spiny trees that were impossible to completely eradicate. When they pitched the tents, there were still sharp little spikes sticking out of the hillside. Along with the incline, it didn't make for much of a camping spot.

By nightfall, Sam Bell had noticed that Marty was not among their ranks. The poor man must have fallen behind somewhere on the trail. In the pitch darkness, Sam ventured back to find him, stumbling by the weak light of a flashlight. About a half mile back, he came upon the laboring man and encouraged him to keep going. Camp was just over the next rise—he could do it. Marty said very little to Sam, and when they summited the last rise and trudged into camp, he fell into the pile of bedrolls and luggage and passed out cold. The others gathered around, worried about their lifeless friend. When someone thought to pass a glass of scotch under his nose, he revived immediately. Marty grabbed the glass and tipped the scotch into his gullet. Another scotch later and Marty had come fully, if not rosily, back to life. His eyes found Guido's. "I want to see you in my tent right now," he ordered. "You too, Sam."

His tent, Sam thought. Where is his tent? But in the darkness Marty found a tent, designated it as his own, and raised the flap for Guido and Sam. Marty had not bothered with the protocol of zipping up the flaps to keep insects and reptiles out, and wasted no time launching into a diatribe he must have been composing all day. Did Guido know what he had done in bringing them all out here? He, Marty, could have died out there today. Did he—or Sam for that matter—have any idea of the liability issues? This could be a legal nightmare for Guido—in fact, Marty could make it one. Had Guido considered for one moment the implications of such a reckless setup? Marty kept venting, no less scathing on his third scotch, which he paused to sip as he glowered at Guido. "This terribly planned little expedition is over," Marty said. "Forget the cloud forest. We're heading back."

Guido listened respectfully and said, "Marty, we can't go back. We're in the middle of nowhere. The driver's gone. There's no phone." Guido let the painful information sink in. "We can't go back," he repeated gently. "We have to go up."

At this point, Sam's eyes were caught by a dark, scurrying movement near the opening of the tent flap. Crawling through the zipper and onto the floor of the tent was a three-inch spider. It headed straight for Marty, who had made his revulsion of insects abundantly clear. Marty had started to talk again and was gaining steam when he glanced down at the spider and, without a word, swatted it with the back of his hand. Eyebrows raised, Guido and Sam watched the spider sail through the air back out the tent flap as Marty carried on with his tirade.

The group was tired and disgruntled, but their moods had improved with cocktails. With another scotch, even dyspeptic Marty settled down. Soon they were all tipsy and happy again. Guido decided to sleep up the trail next to his Mexican friends. They welcomed him by their fire while shaking their heads. "Ay, Guido, you really did it this time." Guido didn't need the reprimand. He was convinced his career was over.

It was not a restful night. In the morning, the Bell party woke up in piles at the bottom of their tents. Some of them were unfit to walk. Guido rearranged the luggage and put the incapacitated on horses, making sure Marty had a mount, which Guido hoped would improve his outlook. Sam and Susie were still on their feet, and welcomed Guido's company as they set off again. Sam was starting to enjoy himself; it wasn't the most comfortable trip, but Guido was bringing the mountain alive for him, pointing out details of the transitioning habitat as they progressed up from the thorny scrub forest into the clouds. It was really quite extraordinary, Sam thought, this mountain in the clouds supporting so much life. For so many species to thrive in such a small space, including a primitive turkey with a horn on its head. He was anxious to see it all. Walking alongside the chickens, however, he noticed there were only a few left. What was happening to them? he wondered.

Guido and the other guides grew silent when they reached a valley where a jaguar had killed some people two years before—just their clothes had been found. The problem with big cats was that they stalked you silently from behind. There was nothing you could do against such a predator; if it was interested in eating you, you didn't have much of a chance. In this jaguar hunting ground, the Bell party chatted away, unaware. After passing through the valley safely, Guido relaxed and joined the conversation again as the party marched up—and up.

Eventually it grew cooler as they moved into the tropical forest. The trees grew far over their heads, blocking out the sun. Great old oaks sagged under the weight of bromeliads, vines, and orchids— each tree a standing garden festooned with scented flowers that grew miraculously from their branches and bark. Their steps fell noiselessly on the forest floor, where patches of sun floated like mirages of light and the ferns grew as big as palm trees. Soon they glimpsed the iridescent wings of a blue morpho, a giant butterfly, as it sailed silently through the trees. The Bell party was finally quiet as they looked around them, seeming to grasp that they were in the middle of a rare and beautiful world.

Guido led them to a clearing at the top of the mountains, where a primitive scientific station had been built to periodically house scientists from around the world who came to study the cloud forest. Here, at least, was a proper toilet. Guido would give them time to enjoy it. Maybe it was the altitude, but it seemed to Guido that his guests were showing signs of recovery. Among the trees and birdsong were sounds of laughter.

As they headed back down the mountain, Guido noticed that the general humor of the group was continuing to improve. Maybe Marty wouldn't sue him after all. As if to challenge this sanguine turn, the afternoon sky marbled with dark clouds. They had just managed to

make camp when the ominous gray ceiling ripped open with a tropical storm. As torrents of rain came lashing down, the Bell party retreated to their tents, staying cocooned long after the rain stopped.

In the cool evening air, the Mexicans built a fire. Guido sat nearby
and listened as they started singing a traditional *canción*. The evening
light was golden, and moisture dripped from the branches of trees.
He allowed himself to enjoy a moment of relief; the storm had passed,
and they had survived the hardest part of the trip. No one had been
bitten by a snake, and no one was seriously ill. In addition, he suspected the Bells had been moved by what they'd seen in the cloud
forest. Sadly, now, in this moment of evening perfection, the Bells
remained zipped in their tents.

The Mexicans had started another song, a soft, lovely melody,
when the Bell party emerged and wandered toward the music. It was
the first time the two groups had joined. The Mexicans widened their
circle around the fire so everyone could sit, and continued to sing.
The Bells joined in on the choruses, and the sweet music swelled and
drifted into the valley.

It would be their last on the mountain. There was one chicken left,
which they would feast on that night. The Bells cheerfully gnawed
away at meat as tough as shoe leather. The next morning, Guido took
a chance. As they headed down the mountain, he found Sam and
without preamble ventured, "Sam, is there any chance I could come
to the James Ford Bell Foundation for funding?"

Sam knew that the trip had nearly gone off the rails, and to say that
Guido was rough around the edges was putting it mildly. The trip
had been terrifically uncomfortable, and his life had most certainly
been in danger, yet Sam had enjoyed himself thoroughly. He had seen
an iridescent butterfly as big as his hand flashing blue through a forest
that was like something out of a dream, with flowers dripping from
trees that looked like they had been there since the beginning of time.
It had been hell getting there, but they had laughed and learned, and
when they'd finally reached the cloud forest, the peace was as thick
and soft as velvet. Sam would remember it for the rest of his life. He
smiled at Guido now on the trail.

"Guido, as a matter of fact I don't think it's going to be a problem at all."

Sam would find Guido's subsequent invitations irresistible. In the following years when Guido extended an invitation to some faraway place, Sam was always ready to pack his bags and go. That Guido did not acknowledge obstacles was a double-edged sword, but Sam found that it made things more interesting.

Guido had cracked a central code to human nature that he himself had not previously understood. People paid a lot for comfort, but safety and sedation were not the ne plus ultra of earthly experiences for everyone. While the Bell trip was 90 percent hardship and misery, Sam and Susie were changed by what they saw, and how they came to see it. They would later remember the near misadventure as one of the high points of their lives—and Guido would win their financial support in perpetuity. Guido had had his initiation into major-donor fundraising, and the trip would come to define his inimitable style.

"The only way to save a place is to get people to love it as much as you do," Guido told me. Our plates were empty and he raised his hand for the check. "And the way to get them to love it is to take them there, and not just for a day—ideally long enough for them to get the same kind of feeling. And it becomes part of them, part of their lives. And if there's a near-death experience, so much the better."

It was soon thereafter that Guido's life hit a sharp bend that changed his course dramatically. While he had been raising money to save distant lands, the rivers and fish he cared about most were slipping into crisis. When new evidence about the declining numbers of salmon in the Pacific Northwest came to light, Guido hung up his blazer and quit his job in D.C. He was needed elsewhere.

CHAPTER 5

SCALING THE IVORY TOWER

I N 1990, WORD OF A disturbing report spread like a virus through the conservation community. The first analysis of Pacific salmon populations had been conducted, and the results were beyond distressing. The great salmon runs of the Pacific Northwest were collapsing. The report sent shock waves through the country. Salmon were disappearing—and they were disappearing fast. In just twenty years, the winter chinook of the Sacramento River fell from roughly 86,000 to 500. In thirty years, the fall chinook of the Snake River had dropped from 30,000 to 1,000. In forty years, Snake River sockeye had gone from 3,000 to one single fish.

Guido had known that the rivers in Oregon were changing; he could sense it in the patterns of the fish. On the Deschutes, there were fewer steelhead than ever before. For some reason, their enduring migration had been altered. Now he knew it was not only his home rivers that were being affected, but the rivers up and down the western seaboard.

The state of Oregon dove headlong into the issue and immediately found itself grappling with all it had failed to understand about one of its most precious creatures. The state's Department of Fish and Wildlife, responsible for managing Oregon's salmon, had been blaming the diminished salmon population on what they believed was the obvious cause—not dams, logging, or agriculture, but overfishing in the ocean and rivers. The agency's response was to build hatcheries, and artificially produce more fish to make up for the lost numbers. Thus the Deschutes and many other rivers in the Pacific Northwest were flooded with hatchery fish, which did not end up being a solution at all. The full picture had become more complicated than the agencies were prepared to acknowledge.

In Guido's mind, it shouldn't have been complicated. Salmon needed three things to survive: access to their home rivers, clean and cold water, and a place to spawn. Oregon had compromised each of these requirements, starting with dams. These concrete walls spanned rivers, blocking salmon from swimming upstream and reaching their spawning grounds. In the Columbia Basin there were at least sixty dams, only some of which had fish ladders (concrete or wooden steps built to the side of a dam that offered fish passage up and over the dam). Salmon, determined to the last, often died in the futile attempt to get past these barriers.

Trees, or lack thereof, presented the next threat. The majority of lumber companies in the Pacific Northwest practiced clear-cutting, razing every tree from a designated acreage and leaving huge squares of denuded land. The removal of trees created a number of problems for the fish. Without their shade, the river was unprotected from the sun and heated up, starving fish of oxygen. The root systems of trees also gave stability to riverbanks, establishing a structural integrity that could withstand the fluctuations brought by the inconstancies of weather and other environmental events, like earthquakes and floods. Without the trees to hold the land in place, the regular winter rains brought these barren, unstable hillsides crashing down into the rivers, turning the water to mud and creating impassable blockages.

Farming and agriculture exacted a final price. Many spawning grounds had been incorporated into farms, which were now barred to salmon by fences and graded land, though some farmers raised the irrigation gates during the salmon migration and flooded their fields with salmon so that the dying fish could fertilize the soil. Throughout the year, farmers used the water from salmon tributaries to irrigate their crops. Eventually, excessive diversions and withdrawals from nearby rivers dried up creeks and streams, while the agricultural herbicides and pesticides leaked into the water system, impacting the aquatic insects the young salmon fed on.

The combination of dams, logging, and agriculture had led to a slow and steady degradation of salmon habitat. Looking back at it all a decade later, it was no wonder that, after millions of years of successful migrations, salmon were finally unable to complete their life cycles. The result was that the most fragile salmon runs—the fish that had the farthest to travel upriver—were starting to disappear. Populations of summer steelhead and spring chinook were the first to vanish. Others would follow.

When Oregon's sockeye salmon were listed as endangered, the state would never be the same. Salmon were a symbol of the region. It was said that they had once swum so thick in the Columbia that one could almost walk across the water on their backs. That the fish were now disappearing was unimaginable, like some kind of terrible magic trick.

The newly written Endangered Species Act had the effect of a bombshell followed by a state of emergency. Suddenly salmon featured regularly on the front page of local newspapers. Something very serious had gone awry in Oregon's rivers, and the industries, practices, and systems influencing critical salmon habitat were scrutinized and assessed for their part in the mess. The list of perpetrators was long, but no one knew exactly where the blame lay. Lumber companies, farmers, dams, private home owners, the state—they were all culpable. Now the ESA brought the federal government straight into the middle of Oregon's biggest environmental crisis to date.

Guido listened to the hubbub. He read every article. Retreating to the Deschutes, he stood in the water, cast his line, and began to think. The system had failed salmon. It wasn't out of lack of knowledge—the needs of salmon were understood—it was that no one had been looking down the road at the long-term consequences of current practices; no one had put the pieces together. What made it unconscionable was that this same scenario had already played out with Atlantic salmon— twice: once in Europe and again on the eastern seaboard. It had been a double tragedy, and it was setting up to repeat itself one last time in the Pacific. After that, there would be no more chances.

The plight of the salmon lodged inside Guido like a shard. It wasn't just the salmon that were at stake. Guido had been raised and nurtured in these same waters; their habitat was his. It was on these rivers that he regenerated and felt the fragmented parts of himself rejoin, giving him the strength he needed to operate in the civilized world. He needed them for his own survival.

Guido felt in his gut that he had a role to play in preventing the impending tragedy, but he was nowhere near the stage. Anyway, who would listen to him? He was a graduate of a mid-level university where he had failed to achieve academic distinction. Up until now, good grades had simply not been important. Up until now he had wanted only to be on the river. But the ESA listing changed everything. The battle cry had been sounded. There was no question that he was joining the war—the question was, at which level? His understanding and knowledge of rivers was unique. He knew he could be more than just a foot soldier now, going door-to-door, collecting signatures. He had seen how it worked in DC: one could progress as far as one's education allowed. The only way for Guido to gain credibility was with a higher degree. He had to go back to school.

It was a bitter pill, but Guido resolved to swallow it. He was motivated now as he'd never been. He suspected that he wasn't an ordinary applicant; he had some unusual abilities and years of experience in the field. With some luck, he could earn a degree that would put him where he needed to be—on the front lines. And in his mind, there was only one school worth trying for.

Yale's School of Forestry and Environmental Studies had fostered visionary conservationists since the early 1900s. Its graduates now affected environmental policy at the highest level. Guido decided this was the school for him. There would, he knew, be challenges to getting in. For starters, he didn't have the academic record he needed to be considered by an Ivy League school. On paper he looked like a derelict who was resistant, uninterested, or possibly mentally deficient. Which meant that, if he was going to have a prayer at being accepted, he would have to break into the admissions process at Yale. He would have to sell himself directly.

It was the spring of 1990 when Guido jumped into his Volkswagen van and got on Interstate 95 heading north from DC, alternately listening to the radio and composing his presentation as he passed through Maryland, Delaware, New Jersey, and New York on the six-hour drive to New Haven, Connecticut. He drove past cities and across bridges that spanned rivers—big, lazy, muddy rivers with only the hardiest grasses lining the banks. This was old-world America; people had been developing industry here for centuries, and it showed. Once Atlantic salmon had filled the waters of the Norwalk, the Pequonnock, the Housatonic, and the Connecticut—rivers that were now so toxic with chemicals that their fish were inedible. The last stretch of highway ran along the coastline, where the Atlantic glittered and recreational boats rocked in their harbors. The fishing boats were long gone.

Set among Yale's centuries-old ivy-covered buildings, Sage Hall was a beautiful old brownstone marked out front by a statue of an intrepid forester outside. The photographs lining the entryway included some of Guido's heroes and teachers. One day, he hoped, he might join them. Guido had done his research; in his pocket was a list of four professors he needed to see. His strategy was simple: he was going to open himself up and talk, just as he'd done presenting at the High Desert Museum and hosting *On the Fly*. He was going to tell them who he was, what he had done with his life so far, and why he belonged at Yale. He scanned the directory and began knocking on doors. By the end of the day he had presented his argument four times, and without exception the professors had listened. Guido's understanding of biological systems was exceptional, and his experience in the field highly unusual. He described the intricate ecosystems of the high desert, the cloud forest, and the salmon rivers with passion and knowledge, speaking about these places as if they were his home, as if he would die to keep them wild.

Guido drove back to DC thrilled to his core. He was sure that, unlike his past teachers, the professors at Yale had his measure. Back in his apartment, he typed a letter to Professor Stephen Kellert, the director of admissions, thanking him for his time and reiterating his profound impression that he and Yale were a perfect fit. A few weeks later, he heard back from Kellert. The committee had unanimously agreed that he was an ideal, if unconventional, candidate and encouraged him to apply.

The next step was less comfortable. Guido winced inwardly when he called to have his college transcript sent from the University of Oregon. While his memory of the details had faded, he knew the document was not a great testimony to academic commitment. It took a few more weeks for Yale to respond. Professor Kellert wrote the letter himself; it was brief and brutal. Guido's transcript was one of the worst they'd ever seen; they couldn't let him in.

Guido read the words without taking them in. Flashing through his mind were bits of evidence from the past that seemed to coalesce into an argument that he did not belong at an academic institution—much less Yale University. While he might have deserved his comeuppance then, everything in him protested Yale's judgment now. They were wrong. They had made a mistake. From thirty thousand feet up, Guido's dispute with the establishment was inconsequential. Critical ecosystems were at risk, ecosystems upon which they all depended. People needed to be educated.

He got back in his van and headed north again, preparing his rebuttal as he drove. College, while not a peak performance, was six years ago, and every year in Guido's life since spoke of his commitment to conservation; every year his understanding of ecosystems grew. He was a willing and able envoy for a vital environmental cause. Was Yale going to slam the door in his face because of a few mediocre grades?

When Kellert agreed to see him, Guido made a simple plea. "Please don't judge me on the past." Kellert remained unswayed. Guido was simply too much of a risk. Guido was silent. There was an obstacle in front of him, a dam in the river he couldn't get around.

Guido had arrived at a reality he had eluded for a long time. To get what he wanted, he would have to play by someone else's rules. He didn't like it, and he might be no good at it, but there was no other way forward. He surrendered and embarked on a plan B, seeking out members of the admissions committee and asking them what was required to make him the most competitive candidate the following year. He wanted to know the rules. The clearest answer came from Kellert. Guido needed to do a few things, none of which was easy. The first was to address his performance in college. He had to explain to Yale, and maybe to himself, how he came by such poor grades. Then he had to demonstrate his passion, discipline, and mission in the real world. He needed to prove he was a man of substance, and that he belonged at Yale. Lastly, he had to brush up on his math skills.

Guido listened to all of this in concentrated silence. "Okay," he said. And then he was gone.

If they wanted bold action, they would get it. Guido bought a ticket to South America and headed straight for the vast rain forest of Brazil. Here, in the heart of the Amazon, the Rainforest Alliance was actively saving the rain forest but had no program for the fish of the mighty Amazon, the river that held the greatest diversity of freshwater fish in the world. Guido had memorized the long list of extraordinary species—the glorious peacock bass, the monstrous arapaima, the silvery arowana, the herbivorous pacu, among many others. He knew that, just like the salmon of the Pacific Northwest, these fish faced threats. In fact, he informed the Alliance, the fish of the Amazon were as important as the rain forest itself. They were an integral part of the ecosystem, supporting the health of the river, which was the central artery of life in one of the planet's most vital ecosystems. Guido reported that there were extensive hydropower developments in the works, as well as mercury pollution from gold mining, and there was the perennial problem of overfishing. Intrigued, the Alliance agreed to consider Guido's recommendations. Guido spent the next six months interviewing scientists and experts in the region, gathering data toward establishing a conservation program. By year's end, he had created a new fish strategy for the alliance that would safeguard the river's fish against these impending threats. The Alliance immediately adopted the program—and then they tried to hire Guido. He declined and took it all back to Yale.

In his next application to Yale, Guido addressed the first two of Kellert's concerns. After describing his activities in the Amazon, he went on to candidly admit that his energies in college had been diverted to collecting reptiles and amphibians, producing weekly segments of a local television show, and continuing his work for the Oregon High Desert Museum. In short, he wrote, he was distracted by his activities in the natural world—and his course work had suffered. But he had never swayed from his true subject, which was what had brought him to Yale. Guido ended his letter with a vow of determination. "I cannot change what I did in college, but I can show a strong record of accomplishment since then. . . . I have been working in conservation all of my life," he added defiantly, "and will continue with or without a degree from Yale."

This time they let him in, but his acceptance was conditional. He had to take remedial math and science classes and demonstrate adequate understanding of these disciplines that were at the core of Yale's environmental program. Guido had jumped through the first hoop, but there would be others. He was beginning to realize that to change the world, he would have to understand it, and this would require facing old fears, and weaknesses he had successfully hidden.

Guido showing his grandfather the day's catch, Deschutes River, Oregon.

Guido in Minnesota getting to
know a new mail-order arrival
(Everglades rat snake).

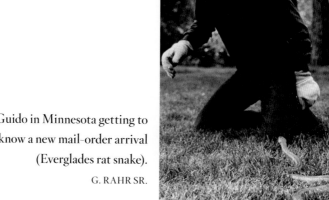

Guido on one of his many solo explorations of the Rahr
property on Tenderfoot Lake, Wisconsin.

G. RAHR SR.

The Rahr cabin in autumn, Deschutes River.

G. RAHR

The entrance to Eagle Creek Canyon, where we rescued
steelhead minnows every year.

T. MALARKEY

Guido and his
younger brother
Willie with a
Burmese python.

G. RAHR SR.

Guido at twenty-four with a Mexican *serpiente tigre*, in the foothills of the Sierra
Madre del Sur, on his way to the cloud forest.

LIBBY JENNINGS

Spencer Beebe, Guido, and Peter Seligmann at the Oregon coast.

WILD SALMON CENTER

Early-morning catch while pram fishing for fall chinook
salmon in the Tillamook Basin.

JEFF STREICH

Peaceful morning, pram fishing in the Tillamook Basin on the Oregon coast.

G. RAHR

Guido with a summer steelhead on the Dean River, British Columbia.

Guido and firstborn son, Gee, fishing on the Deschutes.

Guido in Portland with his second son, Sumner, after a day on the coast fishing for spring chinook.
LEE RAHR

Pete Soverel salmon fishing on the Nushagak River, Alaska. Pete now runs the Conservation Angler and continues to lead steelhead science and angling trips in Kamchatka.

G. RAHR

Brown bear crossing the river at sunset, Kronotsky Volcano, Kamchatka.

IGOR SHPILENOK

Brown bear feasting on spawning sockeye salmon, Kurile Lake, Kamchatka.

IGOR SHPILENOK

Koryak hunter and bear-dog puppy visiting the Wild Salmon Center
camp on the Utkholok River, Kamchatka.

G. RAHR

Helicopter flying over the Krutogorova River, Wild Salmon
Center expedition, 1999, Kamchatka.

G. RAHR

Attempting to dislodge fallen cottonwood across the Krutogorova River, 1999.

G. RAHR

Jack Stanford, Nick Gayeski, and Sergei Pavlov walking a
transect on the Krutogorova River, 1999.

G. RAHR

Dmitri Pavlov, Serge Karpovich, and Ksenya Savvaitova on the Utkholok River, 1998.

G. RAHR

The Wild Salmon Center team meeting in Moscow, 1999. Back row, from left: Guido Rahr, Vladimir Burkanov, Misha Skopets, Pete Soverel, Serge Karpovich, Vladimir Kipriyanov, Serge Maximov. Lower row, from left, Xanthippe Augerot, Ksenya Savvaitova, and Dmitri Pavlov.

WILD SALMON CENTER

Guido and Justice Sandra Day O'Connor after hooking her first steelhead on the Deschutes River, Oregon.

DAVE MOSKOWITZ

Brown bears, a sow and three cubs, Zhupanova River.

G. RAHR

Professor Jack Stanford with a fall steelhead on the Utkholok River, Kamchatka.

G. RAHR

Gordon Moore and guide floating the Zhupanova River with Guido in 2002, Kamchatka.

G. RAHR

Vladimir Burkanov briefing the Wild Salmon Center staff on the poaching
situation in Kamchatka, Deschutes River, Oregon.

G. RAHR

Guido instructing Tom Brokaw in spey casting on the Deschutes River.

JOHN JUDY

Pipeline project, Krutogorova natural gas fields, Kamchatka.

G. RAHR

CHAPTER 6

OF MAPS AND MATH

GUIDO WAS PLEASANTLY SURPRISED that Yale welcomed its incoming students with a wilderness retreat. This was his terrain, after all. He immediately liked the man in charge. Starling Childs was tall and powerfully built, with a handsome, open face and a bright, agile mind. As an adjunct professor of forestry at Yale, Star had the job of taking the newcomers on a four-day excursion into the Great Mountain Forest in northwest Connecticut. Here they lived in cabins in the woods, and the disparate nationalities and disciplines of the first years merged while Star turned their focus to what was growing beneath their feet. As the incoming students breathed in the clean air and blinked in the natural light, Guido looked around at them and believed it possible that some of them had scarcely ventured out of the classroom. As they chatted about their academic backgrounds, Guido briefly visored his eyes to look at a Cooper's hawk circling above them. Star was informing them that the next days would be devoted to creating a map of a single hectare of land. Divided into teams of four, they would take detailed measurements that would describe three dimensions in two, representing the hectare so accurately that an astronaut in space could find it. A map was the quickest and most effective way to communicate information to an audience, Star told them. More than a few of them would be utilizing maps in the future. Guido, like many of the other students, was only half listening.

It was always a challenge to get first-year students to focus. These had been plucked from China, Bhutan, West Africa, and Australia, among other far-flung places, and effectively dropped into a bucket together where they furiously traded information and stories. Star interrupted their socializing with the sobering laws of Pythagoras, explaining that the basic building block of their map would be the triangle, which determined the degree of a rise: the base was the low ground, the short side was the rise, and the hypotenuse was the slope. This was how you made a topographic map. Guido listened unhappily. He appreciated maps; in the past, they had helped him make simple sense of complicated information. But making one? It sounded easy enough, but he knew better. Rosy-cheeked after a dinner lubricated with beer and wine, he approached Star as a fellow naturalist and affably told the professor that it was a great thing Yale had going here and he was enjoying himself in the woods, but, he added, this math stuff really wasn't his thing. In fact he wasn't much good with numbers—not at all.

His confession did not have the desired effect. The next day Guido was assigned to a team that was given a compass and a measuring tape with which they could begin the tedious exercise of determining the coordinates of rise and slope. It would take hours of scrambling through brush, up hills, into groves of trees, down to wetlands, stretching their measuring tape over humps and holes, with two students holding either ends of the tape measure, one jotting down their findings, and one bushwhacking and clearing the obstacles around them.

As Star wandered from group to group, he came upon one team that was missing a member. The three-student team was taking measurements while, on the ground in a nearby gully, Guido overturned rotten logs and wet leaves. When Star inquired about Guido's role in the mapmaking, his team members reported with some pride that Guido had caught a toad and three tree frogs, and had just discovered a population of red efts, a salamander that colonized new bodies of water and hibernated under logs and leaves. Guido informed them that a recent rain had activated the efts to migrate from the uplands to the wetlands. The team was happy to let Guido do his thing, especially as they were learning so much from him. Star noted that Guido had won his group's respect in a matter of hours. He watched with curiosity as Guido lay on his belly in the damp leaves, communing with the efts. He remembered that this was the kid who wanted out of the math. Star recognized an unusual student when he came across one. It would be interesting to see how the exercise progressed for Guido.

That night when the teams sat down and began to transpose their measurements to paper, Star noticed that Guido had again drifted to the outskirts of his group. It was the math, no doubt. Star reminded him that there was more to a map than inverse tangents; their map had to represent exactly what was growing and living on their hectare. They had to distinguish their hectare from all the other hectares. To this end, Star asked Guido what else he'd caught that day. Guido came to life and recited a list of snakes and amphibians, giving a special mention to the red efts, which he had not seen before. This was information his team could use, Star said. Guido considered this and shrugged, agreeing to offer it up.

The fourth day, when the students were tasked with creating their final map, deciding on its aesthetic and scale, Guido again wandered off. Star followed him and suggested that Guido draw the fauna he had observed, providing a visual index of the creatures that lived in their hectare. Guido nodded, accepting the assignment, and went in search of colored pens. Wedging himself among the others working elbow to elbow, he hunched over the map with his pens and began to draw.

Star observed the groups as they employed scientific calculators, rulers, and compasses to faithfully re-create their topographies. Some used the moon as a starting point, with a scale as big as the solar

RED EFT

system. Some burned the edges of their map, making it seem antique. Returning to Guido's group, Star was struck by the transformation of their map, the margins of which were now populated by exquisite renderings of creatures Guido had caught over the days. There was an American bullfrog, a leopard frog, a tree frog, a green frog, a few red efts, an eastern garter snake, a common snapping turtle, and a four-toed salamander. Guido's illustrations brought the map to vivid life, reminding Star of the finest maps of old with their sailing ships, sea monsters, and mythical beasts—maps that delighted and inspired the viewer to imagine the mysteries of the world that lay beyond pure spatial description.

Circulating around the maps that afternoon was also a visiting professor who wanted to write about Yale's student cartography project. Impressed with Guido's team's map, he requested permission to borrow it. After that day, the map was never seen again. Star still mourns its loss. Of the more than six hundred maps he's seen over the decades, Guido's remains one of the most memorable. It would be the first of many maps that would impact the course of Guido's life; in this, Starling Childs had been right. More than any single influence, maps would affect Guido's understanding of the world. They could create new geographies and describe relationships that had been overlooked.

In his first weeks at Yale, Guido met an unusual woman named Xanthippe Augerot, who was exploring just such a geography. Xan was attempting to create a map on a scale far beyond Pythagoras. Guido was instantly fascinated, for at the heart of her map were salmon. Sitting in the lobby of Sage Hall, the two found themselves talking fish. Xan had ventured into a body of water that Guido had never explored, and here she had come upon a geographical relationship few were in a position to see.

Fluent in Russian, Xan had worked on a Soviet fishing vessel in the North Pacific where she had overseen the catch and transfer of fish from American fishermen to Soviet factory boats, alternating between Russian and English on a two-way radio to coordinate with the different parties. It was business on the high seas, and it had given Xan unusual access to a complex intersection of two cultures and systems. In the Pacific, these otherwise antagonistic countries cooperated over their shared resource of salmon. But things had been getting complicated in these northern waters, and it had made her wonder about what in this fluid ecosystem belonged to whom.

Recently, she told Guido, tensions between the Russians and Americans had flared up because salmon stocks were declining in the Pacific. In the chatter on the radio, Xan noticed some vexation among Russian fishermen over new regulations coming from the American fisheries scientists who were responsible for protecting ocean salmon populations. There was no agreement as to exactly what was happening to the disappearing salmon—or where: these fish migrated back and forth across the Pacific Rim.

The scientists could only make guesses, dictating quotas that might or might not protect the balance of the ocean habitat. The fishermen protested this random tweaking of numbers that dramatically affected their livelihood—how much could the scientists really know about what was swimming in the sea or how their regulations were affecting fish in the deep Pacific? There was no way to penetrate the mysteries of such a massive body of water.

But both sides agreed there was cause to worry. The ocean was an ecosystem, governed by the same laws as any ecosystem. When healthy, it was balanced, complete, and self-regulating. When disrupted, it was vulnerable to imbalance, dysfunction, and illness. Something had disrupted the salmon of the North Pacific, but what? And what would happen to the marine ecosystem if a fish that was vital to its health began to disappear?

These questions opened a window onto the complex reality of the ecosystem of the Pacific Rim, a reality Xan committed herself to understanding. By the time she got to Yale, she had earned degrees in oceanography, fisheries management, economics law, and ecology, layers of learning that all went toward solving the puzzle of Pacific salmon, a fish that swam between Russia and America without a thought to national boundaries, linking the disparate countries in ways only a few recognized.

Guido knew Xan's work was relevant to him, and that she was deep into a frontier he was just entering. For years, Xan had been gathering research from fishermen and scientists across the Pacific Rim. Once compiled, this research could create a definitive document of the salmon of the Pacific, a kind of encyclopedic map that showed which salmon were where and when and how many. It would be a resource both scientists and fishermen could agree on and use. But to complete her work, Xan needed money and support. Like Guido, she had come to Yale to get a degree that would give her the credentials she needed to play at a higher level.

Whether or not the two graduate students recognized the connection between their respective missions was unclear. What they knew was that they liked each other—and had a lot to talk about. Xan left Yale after one semester when she realized the program couldn't provide the support she needed. She and Guido would meet again, and one day find a near perfect alignment. For now, all the signs were that Guido was exactly where he was meant to be. But like other things in his life, Yale would not turn out to be so easy.

Guido met his nemesis in the form of Steve Beissinger, a sharp, young professor who taught Conservation Biology, a course designed around a field so new there wasn't even a textbook. When Beissinger saw Guido, he knew he'd met him before. "He's not someone you forget," he later told me. Beissinger had been living in DC researching his postdoc in birds at the National Zoo when Guido had crashed a party at his house. Long after the other guests had gone home, Beissinger found himself trapped with this strange, intense kid who seemed to have wandered in off the street. Indifferent to the hour, Guido interrogated Beissinger, wanting to know everything Beissinger knew about his beloved harpy eagle.

A scientist by training, Beissinger strongly believed in the value of hard science, the core of which was research. There was a lot of *re* to research, he informed his class. To reach conclusions or even suppositions, one had to *re*peat experiments again and again. The results had to be translated into numbers so they could be read and understood—not unlike a map. There was nothing glamorous about it. One had to do things over and over in precise detail until there were enough repetitions to build a pattern. To prepare his students for the research to come, Beissinger started off the year by giving them math-based problem sets, for math was the language they would be speaking.

Guido instantly knew he was in trouble. This research sounded like pure drudgery, and unlike the rest of the students, he was nearly illiterate in higher math. He didn't know how to think about statistics or make basic calculations. It was a muscle group he'd never developed. One day in statistics class, he stopped the proceedings when he asked, "What's a 'log rhythm'?"

When Beissinger took it upon himself to force Guido up to speed, Guido fought back, striking for higher ground by contesting the essential relevancy of Beissinger's mathematical equations to the real issues of conservation. The war to save the planet was on, and Guido doubted it would be won from behind the walls of the ivory tower. While this might have been true, Beissinger agreed, many weapons would be necessary to win such a war. Maybe Guido didn't want what Yale had to offer. If he did, he should get back to work. There was a statistics book in front of him, and course work to finish.

The main event for Conservation Biology was an independent

study project. Guido struggled to come up with a subject he could address. Finally, as his mind wandered back to the rivers of the Pacific Northwest, he decided on a question that had gripped him for years. Why did chinook salmon that were swimming upriver to spawn take bait? It was widely known that as soon as these fish headed for home, they stopped eating. So why could they be caught? Was it curiosity, irritation, territoriality, feeding instinct? He proposed an investigation that would unravel this mystery. Beissinger, not thrilled by a project that would enable Guido's fly-fishing obsession, rejected the proposal. He was beginning to understand what was going on with his most unusual student. Guido's education had holes. His understanding of math was dangerously weak. Somehow he had slipped through the system without having to confront hard science. Beissinger guessed that Guido's skill in the field had distracted from this lack. Perhaps his privilege had also buffered him. Beissinger had seen the name Guido Rahr before, in a display case at the University of Wisconsin. If he remembered correctly, Guido Rahr Sr. was involved with wildlife preservation and owned an enormous tract of wilderness property in the state. His grandson had most likely had it easy in life, finding ways to dodge the system and do exactly what he loved to do.

As the semester continued, it was increasingly clear to Beissinger that Guido needed to be pushed. Avoiding challenges had made him intellectually lazy. More important in Beissinger's mind, Guido didn't yet understand how critical science was to conservation. Without science, Beissinger told Guido, he was just another citizen expressing an opinion. They argued again and again. Guido was frustrated; Beissinger was a naturalist too—why couldn't he understand that experience in the field was more valuable than crunching numbers in a classroom? Beissinger understood perfectly the value of experience in the field, but one couldn't change the world from the field. No matter how skilled you were at catching creatures, Beissinger maintained, you had no chance of protecting them if you couldn't represent them on a scientific level. Without hard data, Beissinger maintained, you had no argument. Observation and collection of data were critical to stating facts, and without facts you wouldn't be able to change policy—if that was what Guido genuinely wanted to do.

Guido had no doubt about what he wanted to do, but he was temporarily trapped, faced with his first major academic obstacle since Ditmars's *The Reptiles of North America*. In fourth grade it had taken all his will and concentration to make the letters face the right direction. Numbers were even less cooperative. Math was a string of nonsensical symbols that he had managed to ignore until now. It turned out one could get around perfectly well without math, and he had done just that for twenty-eight years.

Fueled by anger and frustration, Guido applied himself as he never had to school—even though "school" wasn't the place you necessarily learned the most important things; even though school harbored people with no practical learning. While staying afloat in statistics, he continued to lock horns with Beissinger, though Guido rarely gained a single inch on his professor, who was threatening him with a below-average grade in Conservation Biology.

Beissinger wasn't worried about Guido. He'd watched him in class and seen him interact with his fellow students. Guido had a certain skill with people, a passion that transmitted to others. He was an outlier, but he was also a connector. For all the testiness of their relationship, Guido learned more from Beissinger than any other teacher at Yale. Later he admitted that Beissinger had pushed him to another level.

On weekends Guido left the tedium of high math for the rivers and estuaries of Long Island, where he fished for striped bass and bluefish. The Connecticut coast, however, did not make his heart sing. It was considered a drowned coastline because the sea had flooded the land, submerging the fertile coastal plain. Still, it was better than Conservation Biology, and it was a relief to be in the elements, linked once again by his fishing line.

Every few weeks he took the train down to New York City, where he had been accepted as a fellow of the exclusive Explorers Club. In the tradition of England's Royal Geographical Society, the distinguished club allowed only genuine field explorers to become professional members. To his delight, Guido's expeditions in Chiapas and the Amazon had qualified him. He was entranced by the rooms of the elegant, rambling old brownstone on East Seventieth Street that were populated with mounted polar bears, whale penises, tusks, fossils, and original photographs from expeditions all over the world. He stood for a long time examining a great map stuck with miniature flags marking various explorers' feats, realizing with some amazement that this was his tribe.

Guido made regular visits to the club to attend black-tie dinners and functions, rubbing elbows with noted explorers of his time who had been awarded for first ascents of mountains, expeditions to the north and south poles, the Nile, and even the moon. He also made a point of visiting a polar bear hide hanging on a wall of the third floor. The white fur had a rare translucence in water and was a key component to tying the wings of a coveted fly—the green-butted skunk. Over the next years, a small bald patch grew on the backside of the bear whose fur ended up in Guido's green-butted skunks.

In the beginning of his second year at Yale, Guido made a pivotal find. He discovered the writings of Dr. Livingston Stone, a kindred if bygone spirit. The country's first senior fish culturist, Stone perceived an imminent threat to Pacific salmon more than a hundred years earlier. After watching Atlantic salmon disappear, he saw disaster approaching for their mighty brethren and called out a warning that no one heard.

Stone's vision and boldness spoke to Guido across the ages. Bearded and somber, with clear blue eyes, Stone had ingeniously addressed other declining eastern fish stocks by pioneering the sharing of fish species between the east and west coasts. In 1870, he undertook precarious missions, like transporting 35,000 young striped bass by train over 2,800 miles to California. The fragile fingerlings were carried in wooden barrels and milk cans, and Stone's team cooled them with ice and changed their water every two hours. The containers were agitated day and night by hand as the train navigated rickety bridges, poor rail beds, dry plains, and the Rocky Mountains. Against all odds, the little fish survived the weeks-long journey and went on to thrive in Pacific waters.

After a triumphant return, Stone was called back to California, where he would stage an equally inventive attempt to save the collapsing Atlantic salmon runs. Here he was to collect the roe of the hardiest salmon on earth in the hope that the large and abundant eggs of chinook salmon could populate eastern rivers and lakes. On California's McCloud River, Stone built the first federal fish cultural station, where the roe of chinook salmon was collected and sent back across the country by train. It would take a few tries, but eventually this transplant took, and chinook would adapt to thrive in the landlocked Great Lakes of the eastern United States.

Stone went on from the McCloud to survey the rivers and watersheds farther north, venturing all the way to Alaska. What he saw in the Pacific alarmed him. Even in this vast ecosystem, he felt, the safe places for salmon were limited. People were encroaching on the rivers with their farms and factories, and the waters were filling with fishing boats. The demise of Atlantic salmon was being repeated in the Pacific. "In the Pacific," Stone wrote, "the salmon are about as safe as the fur seals were last year in the Bering Sea." Which was not safe at all.

To Stone, the situation was hopeless—salmon were caught between two forces they could not escape: commercial fishermen and advancing civilization. A witness to the permanently depleted Atlantic salmon runs, he described a devastating corollary: Once a river was degraded to the point of species extinction, there was no hope for either. "All the kings horses and all the kings men could not bring a salmon river back together again." It would take the rest of the world too long to realize what Stone already knew—even hundreds of millions of dollars could not restore such lost habitat.

In the face of this unsolvable dilemma, Stone proposed an idea he thought people would ridicule, as they had ridiculed the idea for a buffalo preserve thirty years earlier. Now the buffalo, which had roamed the West by the millions, were all but extinct. If they had created a protected park for the buffalo, the burly beasts would still be grazing the prairies. The salmon needed something equivalent to this protection—they needed a national park. This neat, efficient, and relatively inexpensive measure could mitigate an otherwise disastrous outcome.

Stone's argument was urgent. The salmon of the United States were one of the country's "most valuable possessions," and "not a day ought to be lost" in finding and securing a park for salmon where they could live and thrive as nature intended. "If we procrastinate and put off our rescuing mission too long," he wrote, "it may be too late to do any good."

Stone's words lit a fire in Guido. He felt in his bones that Stone's strategy was right—he'd seen it in action at Conservation International, where the mission was to protect just 15 percent of the earth's most biodiverse habitat—a reasonable and achievable goal. Could this not be done for salmon? Protecting the healthiest salmon rivers would go a long way toward ensuring the species's survival. The alternative was unimaginable. Salmon would slowly but surely disappear from the Pacific. Guido knew that a world without salmon would lead to a "trophic cascade," the collapse of a vast ecosystem. It had happened with the buffalo. The entire prairie habitat was disrupted and ultimately broken by their extinction. Humans seemed to understand very little about such habitats. Having touched their mystery, Guido intuited that these perfectly balanced systems held secrets worth guarding. He was certain now that, without salmon, the verdant and productive rivers of the Pacific Northwest would change forever, likely endangering the greater habitat of the Pacific Rim.

In his final semester at Yale, Guido wrote the most inspired paper of his life, in which he outlined the basis for what he called a Stronghold Strategy and the value of protecting the last pristine salmon rivers in the Pacific. Instead of trying to mend what was already broken, he argued, protect what was intact. As he wrote, he knew he had found his fight, and it was deeply personal. Strongholds had been with Guido since childhood; the existence of these wild places had allowed him to survive; their very wildness had made him whole.

The trick would be getting others to understand and join him in their protection. But why wouldn't they? They had the knowledge of the past; they had California, and the lessons of the Atlantic. Who could stand by and watch history repeat itself? He realized that most people would do just that. They had to be told again; they had to be reminded. Even if they didn't listen, Guido's mind was made up. Protecting the salmon strongholds of the Pacific Rim was the only way to safeguard a species that was invaluable not just to him but to the entire region. How he would do it, he had no idea.

Guido had only the vaguest idea of how vast the region was. What lay beyond the distant territories of British Columbia and Alaska across the Bering Sea was unknown, but increasingly fishermen talked about a land that had been barred to them for all of modern time. Starting in the late 1990s, faint pulses of information had started circulating among certain anglers. The rumors, hopes, and prayers were all the same—the biggest salmon of all lived on the far side of the Pacific Rim, in the Russian Far East. A place called Kamchatka began to take life inside Guido, for his hunt for big fish was nothing short of an obsession. The more he heard, the more he knew he would give his eyeteeth to get to this forbidden paradise. As the Soviet Union collapsed and the walls came down all around the formidable empire, the doors to Kamchatka remained locked. This was Russia's most closely guarded territory; the peninsula had been a military-only zone since the 1950s. Even Russians could not visit Kamchatka, home to Russia's largest nuclear submarine fleet and myriad top-secret military installations. Security surrounding Kamchatka was beyond tight—it was murderous. When, seven years earlier, a Korean airliner had strayed into Kamchatka airspace, Soviet fighter jets were dispatched to shoot it down. The plane had crashed into the Sea of Okhotsk, killing all 246 passengers aboard. For many reasons, Kamchatka was not like any place on earth. Guido recognized that the peninsula promised discovery and danger in equal measure. As always, the danger failed to register.

One spring day in his last semester at Yale, he was forwarded a ragged letter with a postmark from Moscow. In the tight circle of elite fly fishermen, word had traveled about Guido's quest to find the biggest chinook in the world. The letter, from a Russian named Gennady Zarkov, reported that a 126-pound chinook had been caught in a commercial net on Kamchatka's west coast. This was equal to the largest-ever commercially caught salmon. Also, Zarkov added, Kamchatka's Bolshaya River was said to contain all six species of salmon. Outside Alaska, such species richness was unheard of in the West.

As Guido remembered it, Zarkov's brief report sat inside him like a "stick of dynamite." When the doors to the Russian Far East finally opened, Guido got out the map. Kamchatka was the first major landmass on the other side of the Aleutian Islands. It was a huge peninsula, perhaps bigger than California, but he didn't know the first thing about it.

At Yale was a fellow student named Margaret Williams, who had been working on conservation initiatives in the Russian Far East for the World Wildlife Fund. Guido found Margaret and squeezed her for information. What was Kamchatka like? How could he get there? What were conditions like in the Russian Far East? Did she know anyone who could help him get to the Bolshaya River?

Margaret related that while Russia's doors were open, Kamchatka was a special case. Even with the walls down, Margaret said, access would be hard. There would be visas and permits to get—not to mention the ordeal of simply getting around the place. The entire peninsula had only two partially paved roads—and they connected to nothing. To say that conservation work in the area was a challenge, she added, would be a major understatement.

That said, she knew of one fly fisherman in Russia who might be able to help Guido. Mikhail "Misha" Skopets had been staging solo expeditions in the Russian Far East for decades and was regarded as one of Russia's most brilliant ichthyologists. An eccentric loner, Misha was not easy to find. To raise him, Guido had to pursue all lines of communication. Finally, by fax and phone, Guido wrangled an introduction and, without mincing words, told Misha what he was after. Misha confirmed that there were indeed some big salmon in Russia and that the Bolshaya River was believed to hold all six species, with some char and Dolly Varden thrown in. Guido wrote: "How can I get there—and when?" Misha answered that the time to come would be early summer. Compared to some of the rivers Misha had explored, the Bolshaya was relatively accessible, and Misha would be glad to take him there. Beyond excited, Guido asked his most ardent angling friends if they wanted to join him on what promised to be a phenomenal trip. For one reason or another, the Russian Far East was not a big draw, and he found no takers. This bothered him not at all, and he went forward with his own arrangements, buying a single ticket to Kamchatka, words of warning all but forgotten, visions of giant salmon swimming through his head.

CHAPTER 7

RUSSIA—THE FIRST GLIMPSE

D URING THE SUMMER MONTHS, there is a single flight per day that runs from Anchorage, Alaska, to Kamchatka, Russia. One airline braves the tempestuous course that charts due west along the Aleutian Island range, where the dry, frigid air charging down from the north collides with the wet, tropical currents surging up from Asia. It is a violent and reactive meeting, one that breeds fierce, howling storms that wander swirling and dispossessed across the angry sea.

The first recorded expedition across this treacherous passage was made nearly three hundred years ago by Vitus Bering in his ships, *St. Peter* and *St. Paul*. Bering, along with more than half his crew, did not survive the journey. While the williwaw storms spun their ships in circles for months and scurvy wracked the crew, some went mad and some sank into despondency. One crew member alone seemed unaffected by the brutal conditions, an indefatigable naturalist named Georg Wilhelm Steller. Steller, an intemperate German, had trekked all the way from St. Petersburg to join Bering's expedition, a journey into the unknown that was set to answer, among other questions, whether the Bolshaya Zemlya, the "big land" that lay to the east, was connected to Asia or was a new continent altogether. What they would discover after unimaginable hardship was a mountainous and snow-covered land that would one day be called Alaska.

Commissioned by Russia's Imperial Academy of Sciences, Steller was to collect specimens and record the natural findings of this unknown region. His observations, while gathered sometimes in a matter of hours, are still considered some of the finest, most accurate descriptions of the flora and fauna of the Pacific Rim. In abandoned island caves he found dolls and artifacts that reminded him of the natives in Kamchatka, and he deduced that the two populations had once been joined until a group of them had elected to emigrate over the now submerged Aleutian Island chain. Looking at a topographical map, one can see this land bridge clearly, an oceanic ridge that leads from Alaska's Bristol Bay straight to Kamchatka. It is the same migratory route covered by some Pacific salmon.

Steller's knowledge of the natural world was encyclopedic, and his passion for new discovery bordered on the fanatic. Throughout an expedition that could not have suffered worse luck, he remained intently, maniacally focused on what he might find living and growing on islands that were unfit for human life. He bounded around the *St. Peter* with endless energy, seizing any opportunity to get off the ship and explore the barren terrain. Irascible, opinionated, and often just mean, Steller had little regard for his crewmates, whom he happily abandoned or imperiled as he chased species over hills, dales, and rocky cliffs. Nature alone seemed to sustain him, and this, along with his cantankerous disposition, made him distinctly unlovable.

I realized, when reading about Steller, that Guido was cut from similar cloth. But while Guido shared Steller's monomania for the natural world, he possessed qualities Steller lacked. These qualities would assure Guido's survival just as their absence would seal Steller's tragic fate, dying homeless and friendless in the middle of Siberia. Guido was knowledgeable, but he did not deem his knowledge to be better or worse than anyone else's—it was simply his. In addition, he had a growing willingness to learn from others, as well as an appreciation for realms he had not mastered. He was coming to understand that he wasn't alone, and that he needed others in order to advance his own learning. In Russia, Guido would rely on these capacities as he never had before, for as his plane followed the route of Bering's ships, he was entering territory that was as wild and unpredictable as the williwaw winds of the Aleutians.

Only two roads traverse the 900-mile peninsula of Kamchatka, which juts like a serrated knife into the frigid Sea of Okhotsk, and these have no land-road connection to the rest of the continent. Off the eastern shore the land plunges 27,000 feet into the Kuril Trench, where the Pacific plate is slowly sliding beneath the peninsula, forcing the earth's magma into wondrous steaming and bubbling formations. Geysers, hot springs, and volcanoes dot Kamchatka, and thirty of the peaks are smoking and active. The gods of nature are still at work here, rocking the earth regularly with quakes. A month before Guido arrived in July 1993 there was an earthquake registering 7.5 on the Richter scale—cataclysmic by Western standards.

Kamchatka is, oddly, shaped like a fish, a big, downward-swimming fish that is breaking away from mainland Russia and heading south toward Asia. Running down its center is a mountainous spine, and from these peaks rivers flow to the east on one side of the mountain and to the west on the other. The rivers flowing east merge with the Pacific and the rivers flowing west join the Sea of Okhotsk. Most of the human population lives down by the head of the fish, in Petropavlovsk, the city Vitus Bering founded for its location on the edge of the deep, protected Avacha Bay, where lush green hills cascade down to cerulean water that stretches for eighty-three square miles. With its narrow opening and capacious interior, the bay was a seaman's dream—the perfect place to build and launch ships. Later it provided as good a home for Russia's largest nuclear submarine fleet, a designation that established Kamchatka as a military-only zone.

Though a direct flight from Alaska to Kamchatka takes no more than five hours, the continents are on opposite sides of the international date line, making for a twenty-hour time difference between these neighboring geographies. While a relative stone's throw from Alaska, Kamchatka is eleven time zones from Moscow, which speaks to both Russia's immensity and its surprising closeness.

Misha Skopets was waiting for Guido at the airport, which did not seem like much of an airport at all, with a single potholed runway and no other planes. Russian officials folded into stiff uniforms stood guard on the tarmac, watching impassively as passengers disembarked and were shuttled through customs. Guido was given extra attention for his suspiciously shaped luggage, which included a tubular rod case that looked like it might be holding a weapon. When the case appeared to hold nothing but a fishing rod, the customs officer relented and, unimpressed, let Guido pass.

Misha was wiry and compact, with sharp green-gray eyes and an economy of movement and expression in which nothing was wasted. His scrutiny of Guido was brief. The Westerner stood a few inches higher and his blond hair and blue eyes were bright as little in Russia was bright. The two men had already established a comfort level, for their cryptic communication had revealed that they belonged to the same rare ilk. They were naturalists, but they were also hunters.

As Misha knew of Guido's angling prowess and line-class world records, Guido also knew something of Misha. Described as the Indiana Jones of the Russian Far East, Misha had seen rivers no one had seen, and he had discovered four new species of fish, two of which belonged to a new genus. They knew they had things to learn from each other, and in the coming days they would watch each other closely, like members of the same species that had evolved in different habitats.

After the formalities, they said little as Misha navigated the cratered road into Petropavlovsk and Guido took in the bleak human habitation. The city was gray, with half-constructed buildings and streets that looked as if they'd been shelled. The people were dressed in drab, threadbare clothes and wore wooden expressions. Guido lifted his gaze. Surrounding the city were shimmering, snow-covered volcanoes. From these mountains surged rivers he knew were filled with salmon and what promised to be a lush, pristine ecosystem. The contrast between the human poverty and the biological wealth was jarring.

Misha explained that they would stay the night in Petropavlovsk. The next day they would do their shopping and collect their gear. They stayed at an apartment that belonged to one of Misha's friends, lugging Guido's bags up a bare concrete stairwell with broken light-bulbs and crumbling walls. To Guido, everything seemed broken and there was an air of uncertainty about the place. Over a simple dinner, Misha described why: When the Soviet Union collapsed, so did cities like Petropavlovsk. In early 1992, the plug was quite literally pulled from the Russian Far East. Overnight, jobs were taken away, utilities turned off. In winter, people froze without electricity, while others starved when the great machine that fed the people was unceremoniously shut down. Under the Soviet system, people had been provided for; there had been jobs and subsidies. State agricultural farms and hothouses grew vegetables to sustain people through the long winters; dairies made yogurt and cheese, and fisheries caught and processed salmon. There were trucks that delivered these goods to stores. Without this support from Moscow, life was barely possible in this outpost at the edge of the Russian empire.

They called it *razvalilsya,* or the ruining. Everything was difficult now. Where there had been a militaristic order, now there was chaos. The stores were empty of goods. Most of the population had to grow their own food in vegetable gardens. Those with education were forced to turn to physical labor. Misha's family of doctors and musicians procured the majority of their food by fishing, hunting, and gathering. With the disintegration of social order, crime had risen. Everything and everyone was for sale. The nation had gone rogue. Misha said it was natural that in the absence of higher laws, the law of the jungle prevailed.

Salmon were an immediate target for exploitation. Their meat and roe were invaluable both in local shops and on the black market. Now that there was no one to enforce regulations, fishermen stood at river mouths and effortlessly netted as many fish as they could transport, hauling thousands of salmon from the river and dumping them into the backs of trucks. Farther afield, poachers found ways to access wild and roadless rivers where countless salmon were there for the taking. Commandeering old Soviet armored personnel carriers, poachers flattened paths through the untrammeled wilderness. Once on the river, they would set up river-spanning nets to trap migrating salmon, gutting the females for their eggs and tossing males and females alike on the riverbank to die. Such poachers killed tens of thousands of fish in a matter of days. The salmon roe was salted, stored in casks, and buried. Later, it would be unearthed and transported by helicopter to illegal markets, where it was sold by the ounce, like a narcotic, to buyers in Russia, Japan, and China. Hundreds of tons of roe yielded billions of dollars in revenue.

No doubt the surge in poaching was taking a toll on the salmon population, but no one was looking out for the fish anymore. With all of its militaristic restrictions, the Soviet state had protected salmon, regulating catch and monitoring salmon runs and individual fish for health. Soviet fisheries had been well-run operations that caught, processed, and distributed a controlled number of fish. Salmon had thrived under Soviet rule, but now it was open season, and everyone wanted fish. Including Westerners like Guido, Misha observed with a wry smile. The good news was that with the current dollar-to-ruble exchange rate, they would be able to travel in style. For $50 a day Guido could have a personal guide and a cook. Guido wasn't sure what to say about this—he had never traveled in such luxury before.

In Petropavlovsk, Guido was quickly exposed as a foreigner. Russians who learned he was from America stared like he was an animal in a zoo, and sometimes reached out tentatively for a handshake. Some of them burst out laughing as if overcome by the cognitive dissonance of seeing a Westerner in their far-flung city. What had brought him to Petropavlovsk? Guido told them he had come to fish, an answer that elicited bewildered stares. Fish? the Russians inquired. Whatever for? Fishing was a necessity, not a pastime; fish were caught to eat—why would someone come halfway around the world to catch and eat Russian fish?

Misha translated the conversations with faint humor and the detachment of someone who lived apart from society, about which he seemed largely incurious. Nor, Guido would find, did Misha need much from other people. He had lived his adult life in the wilderness. This was where his field of inquiry lay, not in the complex and often meaningless subtleties of human interaction. Guido understood perfectly.

I can imagine how it was for Misha and Guido to find each other and recognize the unique set of traits and abilities they shared. While they did not speak much of the same language, they grasped each other's meaning well enough. There was no need for chitchat; silence suited them both fine. When they did talk, they found some surprising shared ground. Misha's favorite literature overlapped with the few books Guido had been eager to read. They had both devoured Jack London as well as Vladimir Arsenyev's *Dersu Uzala* and sought out the biographies and diaries of other explorers of the Pacific coast and eastern Asia.

They spoke at length about their favorite, *Dersu Uzala*, Vladimir Arsenyev's telling of his explorations of the vast taiga region in the Russian Far East. One night after dinner, Arsenyev's exploring party was visited by a solitary man who stepped into the firelight wearing a deerskin jacket and holding a rifle. He announced that his name was Dersu Uzala and that he was a hunter. When asked where his home was, Dersu responded that "a hunter has no home." He had the high cheekbones and almond-shaped eyes of a Mongolian, though his hair was a tousled blond. Arsenyev was most struck by his eyes, which were dark gray, steady, and spoke of "resolute power." The two men talked late into the night. Dersu told Arsenyev that he had lived his entire life in the open, and Arsenyev would discover that Dersu could read the land like no human he knew. In the broken twigs and disturbed soil of the forest, Dersu could see what creatures had passed, and when. After losing his entire family to smallpox, Dersu had lived as a primeval hunter, exchanging his goods with the Chinese for gunpowder and lead.

Dersu stayed with Arsenyev in the following days, teaching him to read the signs in leaves and grass and branches and occasionally saving his life. They would go on to form a singular friendship in the coming years as they explored new wild and uncharted regions of the Russian Far East.

Years later, when Guido was first telling me about Misha and the many things he'd learned from him, he described him finally, with a chuckle, as his Dersu Uzala.

Ever thrifty, Misha had found an unexceptional guide named Nicolai, a seventy-year-old man with pale blue eyes and white flyaway hair. Nicolai had an old military jeep, and they both looked like they'd seen better days. Guido watched as Misha and Nicolai crammed the jeep with two inflatable rafts, oars, tents, sleeping bags, food, a single pot for cooking, and fishing rods. After squeezing themselves in with the gear, they were off.

The city limits quickly gave way to the immense wilderness of Kamchatka. They bounced and veered along a rough road through forests of birch. In the distance to the east lay mountains and broad valleys rounded by glaciers. As far as the eye could see were windswept forests of stone birch, poplar, and larch. Guido found it difficult to concentrate on the scenery, as it seemed Nicolai was struggling to control the jeep. Something was clearly wrong with the steering. Guido tried to catch Misha's eye, but Misha stared straight ahead, as if this was business as usual. He wore the same impervious expression Guido had seen on the people in the streets of Petropavlovsk. This somewhat aggressive neutrality, Guido would learn, was a protective coping mechanism deployed by Russians when circumstances were uncontrollably bad. Because circumstances were often uncontrollably bad, this was a useful adaptation. If something "inconvenient" was going to happen, well, it was going to happen. There was nothing you could do about it—there was certainly no need for panic. Panic was a waste of energy. As evidenced in Misha's nonchalance while riding in a jeep with no steering, the most effective response to a potentially unpleasant situation was to shut down operations and stare into space.

Guido possessed no such mechanism. To distract himself from the death ride, he made conversation with Misha, asking him about fly-fishing in Russia. Misha roused himself from his self-induced trance and answered laconically that there was no fly-fishing in Russia. He had taught himself to fly-fish because, in his opinion, gear fishing was artless and without challenge. On this the two men agreed. It hadn't been easy for Misha to learn, because no one else in the region knew the sport. He'd searched high and low for information, finally unearthing a few books from the time of the czars and a small brochure printed during the Communist period. From this paucity of sources, he'd surmised the basic principles and made up the rest.

With no fly-fishing tackle in the stores, Misha fashioned his own rods and reels out of deconstructed gear tackle. For fishing line, he used whatever was available—first a clothesline, then one made from the PVC cover that he pulled off electrical wire in pieces and threaded together with nylon line and glue. For flies, he collected fur and feathers from local hunters he knew. He built his own tying tools and vise—and he also devised his own tent and sleeping bag, as well as a lightweight inflatable boat.

Misha had learned a significant amount in isolation and could hold his own as an angler, especially now that better equipment had found its way from Eastern Europe. But it was nothing compared to Guido's gear. Misha was looking forward to seeing the graphite rod freed from its gleaming aluminum tube, the reels from their soft leather cases, and the orderly fly boxes with their many tiny compartments. And to seeing it all in use by his expert companion. "Tell me," Misha asked, "what other salmon species have you caught on a fly?" Guido answered he had caught chum, pink, coho, chinook, and sockeye, as well as steelhead.

"But not cherry," Misha said.

"Not cherry."

"We'll see what we can do about that. Maybe you could be the first person to catch all six species on a fly."

While Guido had his heart set on a monster chinook, he accepted this as a worthy second goal. The land opened up around them, and stretching in all directions were fields carpeted with high wild grasses sprinkled with multicolored flowers. Misha pointed out ferns, nettle,

cow parsnip, and a white flowering plant he called Kamchatka mead-
owsweet. This luxuriant offering constituted a fast-growing annual
floodplain vegetation known in Kamchatka as *shelomainik* that came
bursting forth every spring and remained lush into late summer.

The spare volcanic geology of Kamchatka is marked by the continuous exchange of water as it transmutes from land to sea to sky. In
winter, heavy snows blanket the peninsula's mountainous spine. In
spring the snow melts, collects, and races down gulches and ravines.
Once out of the mountains, these tributaries join together, forming
small rivers that grow larger as they meet and marry and flow ever
stronger to the floodplains. What happened in these floodplains,
Guido was curious to see. He had never seen an untouched floodplain. In the American West, these flatlands were the first acres the
settlers had planted. Fertile from decaying salmon, crops sprung from
the soil and grew tall and strong. Unbeknownst to the settlers, they
were planting on top of salmon spawning grounds.

The Bolshaya, located near the top of the fish's head, is formed by
the confluence of two large rivers, the Plotnikova and the Bystraya,
that come cascading out of the central mountain range and flow west
into the Sea of Okhotsk. The plan was to put their rafts in high on the
northern fork, the Bystraya, a name that translated to "fast." From
there they would float eighty miles to the mouth of the Bolshaya.

For Guido, the Bolshaya had assumed mythic significance. It was
about 1740 in this very river that Georg Wilhelm Steller had made
the first collections of Pacific salmon to reach the scientific community. Sensitive to provenance, Steller used the local Russian and
native Kamchadal names to describe the new species. Thus the Kamchadal name *tshawytscha* was given to the chinook, *kisutch* to coho;
the Russian word *gorbuscha,* meaning "humped," was given to pink
salmon, and the Kamchadal name *keta* was given to chum. These
names remain the formal scientific names for the species today. Steller
had been astounded by the fish in Kamchatka's rivers, which, he
wrote, were "full of rarities and almost unbelievable circumstances."

The abundance and diversity of life inspired an unusual reverence in the scientific Steller. He wrote that "the merciful love of the Almighty is clearly mirrored and revealed for all the world to see through these creatures."

As the jeep bombed through the tundra on a gravel road, the summer air blew in through the windows, warm and humid. Guido tried to ignore the fact that the vehicle's steering wheel seemed virtually disconnected from the rest of the car. They continually veered off the road in slow, uncontrolled slides. Sixty miles later they miraculously reached the put-in. Nicolai set out some bread and hunks of sausage and cheese and a bottle of vodka for lunch. The old man chased shots of vodka with raw cloves of garlic, which he munched on like a bovine. Guido realized Nicolai was drunk, and he was going to stay drunk.

When they unpacked the gear, Guido had a chance to inspect the rafts, which did not look at all adequate. In fact, they looked cheap and dangerous, like glorified inner tubes. Misha and Nicolai were stacking them high with gear. The oars were plastic and dinky and the oar locks were no good. Peering into the contents of the rafts, Guido could see no life jackets.

When he inquired, Misha shrugged in response. "We should be more concerned about bears."

Guido had heard about the enormous bears of Kamchatka. He assumed that, like in Alaska, they would have precautionary bear spray, or even a gun, but Guido saw neither of these. He ventured another question. "What do we have to protect ourselves against bears?"

Misha held up his little filet knife. Guido stared at the knife. "Seriously?"

"Maybe there won't be so many bears," Misha conceded. "I don't know. I've never floated this river." At this, Guido looked the other direction. He was beginning to understand why even his most lunatic fishermen friends had been unwilling to risk the Russian Far East.

Misha was perplexed by the American's obsession with safety. Like most Russians, he was fatalistic. Life in Russia was unpredictable; it could be taken from you any day of the week by natural forces, by the government, or—these days—by criminals. Safety in such a place was an afterthought. Guido did not share Misha's sang-froid. It seemed that between the lousy gear, the unknown river, and the gigantic bears, they were being offered multiple and idiotic ways to perish. As he focused on the water in front of him, his eye was caught by glinting movement beneath the surface. Looking more closely, he saw they were baby salmon, char, and trout parr, all dart-ing around his legs. The river was filled with baby fish. Issues of safety aside, Guido was exactly where he wanted to be.

They finished packing the rafts and shoved the precarious arrange-ments into the river, where they were swept swiftly into the current. Misha had donned his somewhat unorthodox fishing uniform of camouflaged waders and a paratrooper's leather helmet from the last world war. The cap fit snugly over his head. From under the brim of his own worn fishing hat, Guido studied the landscape. There was no sign of civilization anywhere—not so much as a distant contrail over-head. Just endless forests and the unbroken sky. The scented air hummed with insects and birdsong.

The Bystraya, true to its name, ran fast and shallow, making its way to the sea in sweeping curves. Periodically, Misha pulled their raft over and they waded into the cold, clear current to cast their flies. On the shore, bear tracks abounded and Guido tried not to be dis-tracted by the oversized paw prints pressed into the soft soil of the riverbank. In the early afternoon, an expansive hatch of brown may-flies fluttered above the water in a cloud, rousing schools of rainbow trout and grayling that looked, to Guido, relatively enormous. He watched as the fish fed voraciously on the mayflies. Then the two of them created a new fly that imitated the salmon parr the trout were feeding on. They called it *mykija,* and it had a white strip of rabbit fur for its back, a body of translucent yarn, gray hackle, and nickel eyes. Casting the fly, Guido almost immediately hooked a hard-fighting rainbow that was one of the biggest trout he'd ever caught. Misha nodded approvingly and told him to keep it for dinner.

They floated into the afternoon as the sun mellowed. Along the riverbanks, reeds and grasses flickered with life. Swarms of insects hovered in clusters, and butterflies flitted through the grasses, the sun catching their wings. Stalking the shallows on their long stick legs were herons, and in the still side channels mergansers bobbed and fished. Guido spotted loons paddling in pools and, every mile or so, a Steller's sea eagle perched high in a tree or flew overhead. These raptors were larger than bald eagles, with white shoulders and tail and a bright yellow beak the size of a fist.

That night Guido offered to cook his trout, but Misha insisted that he would make dinner—Guido was the guest. Guido was happy to have Misha cook. Anglers had their own timeworn recipes and cooking techniques, and Guido was interested to see what Misha had in his repertoire. On the cutting board, Guido expertly filleted the fish, separating the head, tail, and spine. When he presented Misha with his two perfect fillets, Misha thanked him, took the cutting board, and dumped the entire contents into a boiling pot of water. Then he added some salt.

Guido stared into the pot as the head, tail, and body circulated in the roiling soup. "Is there any part you don't eat?" he asked.

"Perhaps," Misha mused, "the lower jaw. The heads are very good for you," he added. "They're full of vitamins."

Misha divulged that his second marriage had been to an indigenous woman, whose people had for thousands of years subsisted on salmon and knew all the culinary preparations. As a primary food staple, salmon were annually harvested and dried, smoked, and fermented to last the winter. Every part of the fish was utilized: the heads were fermented and used medicinally; their skin was fashioned into shoes. It seemed that the native people of Russia hadn't fared any better under the white man's rule than Native Americans had. The Soviet system had forced tribes across the Russian Far East to abandon their remote hamlets and gather in settlements, where they became laborers on farm or fishing collectives. Like Native Americans, these tribal people would never recover their independence, or the purity of their relationship to their land. Misha added that his first wife's mother had been in the gulag in the Russian Far East—and that everyone in this part of the world had known someone in the gulag at one time or another.

After dinner, they sat by the fire and talked into the night. Like Guido, Misha was the black sheep of his family. He was only sixteen when he knew he wanted to be an explorer in the field of fish science. Raised in the city of Yekaterinburg, Misha had a life that was safe and relatively secure. He had family and friends, a good house, and job possibilities. It was a future that held few surprises. Misha knew the rivers he would be fishing—he even knew the species of fish he would catch. The predictable course that lay before him set him to dreaming of a wilder life, like those described in the books he read by explorers of the far east and north, whose lyrical renderings of these vast, uninhabited expanses made Misha's blood boil. He was powerfully and, ultimately, irresistibly drawn to the mystery and challenge of the wilderness. With all of its uncertainty and isolation, this was a life that he could feel.

At twenty, Misha relocated 5,300 miles east to Magadan, a remote city that sat on the eastern edge of the Russian mainland, overlooking the ice-cold Sea of Okhotsk. Misha had resolved to travel into the unknown—as far north and east as possible. It was, he believed, the smartest decision he'd ever made. After earning his PhD in ichthyology, the prestigious Russian Academy of Sciences hired him to explore the farthest reaches of the country, where there were no fish biologists, and no one conducting science. The academy supplied Misha with whatever he needed in the way of transportation and supplies. Traveling mostly by helicopter, he was able to access any river in the Russian Far East—and there were hundreds of them. For two decades, Misha had the rivers of the far east to himself where, like Dersu Uzala, he hunted alone. He established his value early on. If it was out there, he could catch it. And as with Guido, catching was Misha's passion.

As the moon rose, Misha told Guido about some of the strange fish he had seen. There were anomalous salmon and char, with all sorts of curious mutations and behaviors. He'd found a char with extraordi-

nary yellow lips that was later described as a new species, Levani-
dov's char. His greatest discovery had been in a lake in Siberia, where
a meteor had struck and left a cavernous impression. The water that
filled the crater was deep and cold and covered with ice for all but a
brief time in summer. Once, when dissecting the contents of another
large char here, he'd found the skeleton of a smaller char that had
skull bones that Misha had never observed. Misha waited until the
ice melted again and dropped his gill net hundreds of meters down
into the darkness to see what he would find. From the depths, he
pulled up a fish he had never encountered before; it looked part char
and part grayling. The fish was later determined to belong to a new
genus of salmonids, and given the name *Salvethymus*. Then, in the
Nagaev Bay of the Sea of Okhotsk, Misha was looking under the
rocks at low tide when he found a tiny, long fish that looked something like an eel. Having never seen such a creature before, he sent
samples back to the lab, where no one else could identify the fish either. Misha had discovered not just a new species, but a new genus. It
would be named *Magadania skopetsi,* after Misha.

Guido listened transfixed as Misha described how he would study
the map, choose a river, and, when the snow melted, get in a helicopter and go. Dropped in the middle of nowhere, often hundreds of
miles from civilization, he would live alone for a month or more,
sleeping in his homemade tent and sleeping bag and maneuvering
around rivers on his little homemade raft, collecting samples and
catching fish. Back at his camp, he would measure, weigh, and dissect
the fish, scraping scales and extracting the tiny otolith bone in the
center of the fish's head.

This kind of fish science, morphology, was the backbone of Russian ichthyology, and it focused on the anatomical design of the fish,
which could tell you almost anything. The scales, fins, and body
shape indicated where the fish lived, what it preyed on, how deep and
fast it swam. Not all fish had scales, for instance. Some had large,
bony plates, and some had none at all. Active river fish like salmon
and trout had many fine scales. Swimming ability was revealed in the
shape of the tail fin. Quarter-moon-shaped tails belonged to fast-

swimming fish, while round or square tails were attached to big, bottom fish. The black box was the otolith bone, a tiny calcified plate in the center of the fish's head. The otolith was made up of thin, onion-like layers, and in these layers was written the life history of the fish, as detailed as a diary.

Misha found it miraculous that such a trove of information was stored at such a microscopic level. In a sliver of bone smaller than the tip of your smallest fingernail, one could determine age, ripeness, sex, fecundity, stomach contents, the presence of parasites—even the amount of fat in the intestines. Misha had found that a polished slice of the otolith bone revealed thin layers of clear and opaque bone that represented the difference between the salmon's growth during the day and night. Most important, one could see how long a salmon had been in freshwater and how long at sea, which opened a window onto a previously impenetrable mystery. For two decades, Misha built collections, gathering everything he thought was a species and preserving the more unusual fish skulls in salt solution for the scientists back at the lab.

The problem with morphology, in Guido's mind, was that such biological analysis required killing the fish. Guido informed Misha that in the West, scientists used scales and fin clippings for DNA analysis, which did not hurt the creatures. Misha absorbed this detail with vague interest. He was used to killing fish and found nothing wrong with it. Even with poaching, there were still plenty of fish. That might not last, Guido said.

The days passed quickly as the men settled into the rhythm of the river. They rose in the gray dawn and drank instant coffee with water boiled over Nicolai's cooking fire. The long daylight hours were spent on the water, fishing and floating down the river. They pulled ashore at lunchtime to eat cheese, sausage, and brown bread. Sometimes Nicolai boiled noodles and added *cheremsha,* a garlicky herb that grew among the streamside grasses.

Misha had been studying Guido's tackle, which was without question superior. His graphite rod was strong and flexible, and his reel was smooth and silent. He watched as Guido observed the river, decided on a strategy, and chose a fly. He watched him quickly tie complicated knots between a succession of monofilament lines, all with different lengths and densities. They were connected with a confluence that prevented the line from breaking, even when a big fish was on the line. Misha noted the long, graceful back-and-forth motion of Guido's cast, and how artfully he set the hook in a sizable chinook.

Misha took particular interest in how Guido fought the fish, letting it run and reeling it in, letting it run again. When the chinook finally tired itself out and was reeled in, Guido reached for his net and knelt in the water, where he held the net underneath the fish, raising the handle so the salmon was confined but still submerged in water. This, Guido explained, was to minimize the trauma to the fish. If you took the fish out of the water to measure and weigh it, you had to make sure to return it to the current every minute or two and hold it upstream so it could breathe and revive itself with oxygen.

This was all news to Misha. It was a great surprise when Guido removed the hook altogether and let the fish swim away. Releasing fish was not a Russian practice. They did this in America, Guido explained, because they didn't have that many fish. Salmon always had somewhere to go, and it was important to let them get there—especially during spawning time. If salmon couldn't spawn, there would be no new fish. Misha of course understood this, but there were so many fish in these rivers it hadn't occurred to him that killing a few would impact their numbers—but that was before the poaching epidemic. Who knew how many salmon were being trapped in poachers' nets. Guido then showed Misha how to catch a fish without injuring it by pinching the barb flat with pliers. Without a barb, the hook was easy to remove and didn't tear up the fish's mouth.

This made it harder to keep the fish on, Misha observed. It made it a fair fight, Guido said.

The next afternoon Misha rowed them to a side channel where he had his eye on a riffle. Advising Guido to tie a small fly, they cast on the sunlit water. Soon they felt little pulls that might or might not have been fish. Misha was the first to hook one of the dainty nibblers. When he pulled it from the river he gave Guido his first look at a cherry salmon, the rarest of the salmon species. Guido immediately went back to casting, boring his focus into the riffle, and hooked one soon thereafter. The cherry fought well for its size, and after reeling it in, Guido saw that the spirited salmon was shaped like a small chinook. It was just turning to its spawning colors, with silvery gray scales banded with red.

That night Misha poured two shots of vodka and toasted to Guido, who was surely the first fisherman to catch all six species of Pacific salmon on a fly. Later, as the stars wheeled overhead, Misha shared with him tales of the most elusive fish of all—a fish that even he hadn't caught. It was, he said, a giant trout, rumored to be as old as the dinosaurs. Guido knew these ancient trout; they were called taimen, and he had fished for them in Mongolia. In Russia, Misha told him, they were even bigger—growing up to a hundred or even two hundred pounds. A few rivers in the far east were said to hold these monster fish, called Siberian taimen. Misha had never seen one, but he was definitely looking.

As the Bolshaya swept them closer to the freezing Sea of Okhotsk, the air grew colder. The next morning they rounded a bend and came upon a large yellow raft filled with a drunken off-duty Russian submarine crew. They were fishing Russian-style, with clunky, homemade spinning rods. The lures were large and weighty, and Guido watched as they heaved them out across the water, shattering the surface like rocks.

Guido and Misha pulled over to ask about the fishing, but the men were guarded and eyed Misha warily, unable to place the Russian in camo waders and a leather aviator's cap. When Guido asked something in English, the submarine crew averted their collective gaze and stared at Guido like he was a hallucination. One of them had a revelation and muttered, "Ah . . . Americanski."

To ease the tension, Guido smiled and offered to demonstrate flyfishing, which none of them had ever seen. They watched in amazement as Guido hooked a sizable char on what looked like a little piece of fluff. The Russian crew cheered and applauded and, though it was nine-thirty in the morning, insisted on toasting the astonishing feat with vodka. Soon the two groups were singing and laughing. Before they left, the Russians insisted that Guido join them in a toast to the Russian nuclear submarine fleet.

Misha commented as they left that drinking was part of the fishing experience here, that vodka had probably saved a lot of fish in Russia, as most fishermen were passed out by noon. With a chuckle, Misha suggested that this was Russian "conservation."

On the fifth day, the Bolshaya River was joined by the Plotnikova, and the two small rafts found themselves in a big, fast body of water. Guido watched as they were presented with a labyrinth of braids that wound through strands of larch and alder amid the soft gravel of the Bolshaya floodplain. The river began to divide, and divide again. There was an alarming amount of wood floating around them, which bumped against the shore, snagged on underwater obstacles, and collected into logjams. This was a real concern. If they got sucked into a logjam, their little rafts would go under—and they would stay under. Guido watched as Misha chose one channel after the next, focused and calm and in his element. Guido had no idea how to navigate such a complex river. He had never seen anything like it. What lay before him was upending his perception of what a river was.

The next day, the mouth of the Bolshaya widened and joined the cold gray sea. Miles and miles out in the cloud-covered water were untold numbers of fish that flowed to and from Kamchatka's rivers, which Guido now believed to be the most miraculous watersheds on earth.

CHAPTER 8

OREGON TROUT

WHEN GUIDO RETURNED FROM KAMCHATKA he started looking for a map that didn't seem to exist. He searched atlases and travel stores but could not find a map of the Pacific that showed the ocean in its entirety. Most maps showed either the Russian Far East or the American West, both with only a portion of the Pacific Ocean. Finally, in a used-book store, he came upon an old oversized map that showed the whole ocean, with Russia and America on either end, reaching across the water like two hands. He stared at it for a long time, certain that he was looking at Xan's new geography, one that represented relationships that had gone unrecognized.

Guido and I were both living in Portland then, emerging from an annus horribilis that had seen the premature death of my father and our grandparents shortly thereafter. These were our wise elders, individuals who had offered a spirited gamut of instruction. Gone was the brilliant but brutal egocentricity of our grandfather, the wit and wisdom of our grandmother (who had lived the second half of her life as an Episcopal nun), and the deep, quiet integrity of my father, their cherished firstborn, who seemed to exist in a place above human skirmishes and was dear to Guido. We buried his ashes at the Deschutes because it was the place he had been happiest, the place we were all happiest.

The managing of the river cabins had fallen to us. Early into our tenure, we had a sobering exposure to the changing climate, and, for the first time in our family's history, the Deschutes River flooded. Record rainfall and snowmelt combined with unseasonably warm temperatures raised the water table statewide, and in early spring, rivers jumped their boundaries. We watched as the Deschutes rose from the deep, worn path it had carved in the valley, gaining inches, and then feet. In a few short days it was spilling over the uppermost banks and making its way up the gentle slope of grasses and trees that led to our cabins, dissolving rock, soil, and vegetation into a roiling crush of silt and mud. The oasis our grandfather planted was submerged; flowers and bushes and grass were swallowed; boulders were lifted and rolled away. Our cabins were filled with three feet of water, destroying furniture and rugs and seeping into the walls, where we would smell its dampness for years.

The flood changed the Deschutes forever; Frieda's Riffle was no longer a riffle, and the deep, tugging currents in Master's Eddy were now unpredictable. The spawning beds at the mouth of Eagle Creek were blown out with high water. These changes shook our sense of permanence, altering something we thought was inalterable.

Later that year nature continued its declaration of change as a forest fire raged across the hills and swept the grassy plains, jumping from sagebrush to sagebrush and igniting juniper trees like Roman candles. The Warm Springs Tribe fought it with their helicopter, dipping a large canvas container into the river and dumping it along the containment line. From where we stood, it looked like a perfume puff of moisture that couldn't possibly temper the flames below. Anyway, the land was ready to burn and there was little anyone could do to stop it. We stood at the ready with our hoses in case the fire was blown east, toward us. At night we watched as the bright orange line advanced or retreated with the wind.

The fish, of course, survived, but they had changed too. Guido told me he no longer saw the wild steelhead of Eagle Creek that he had caught in the eighties. Every autumn these fish had gathered at the creek's mouth preparing to spawn. When the fall rains came, and water from the hills trickled down and joined rising groundwater, the creek swelled to capacity, allowing the steelhead to swim upstream to spawn.

These were the fish our grandfather had taught us to rescue every season when the creek ran dry. By early summer thousands of minnows were stranded in warming pools. We played their terrifying saviors, our shadows looming over them on the banks of the creek as they fled our nets, flashing quick and bright in the clear water. Even in the warming, oxygen-thin creek, the little steelhead were incredibly hard to catch. They were survivors, Guido said, like their parents.

Years later I understood that the steelhead of Eagle Creek were special. They were on a timeline more urgent and precise than most salmonids. The creek's smaller, warmer water meant that baby fish hatched earlier and grew quickly in the protected canyon. For a fish, it was a fine place to rear. But the parr of Eagle Creek could not languish. Their rapid growth was linked to an equally rapid timeline dictated by the coming of summer, and a creek that would eventually run dry. As soon as they were able, the little fish had to make their way downstream to the Deschutes. In this big, new water, they would continue to grow until they headed downstream once more to the sea.

When they came home years later, the steelhead of Eagle Creek had to manage the timing of their migration, coordinating their own sexual maturity with the particular changing of the seasons of their home creek. It was a race that began when they were far away in the ocean and felt the first instinct to return home. The steelhead hen's eggs were only microscopic dots, but they would develop as the fish swam. By the time the Eagle Creek hens reached the mouth of the Columbia, their eggs were the size of BBs. From here, the hens would face seals, commercial fishermen with nets, recreational fishermen casting flies and lures, the daunting Bonneville Dam, the fifteen-foot drop of Shers-ars Falls, and a host of predators that lived for the annual fish migrations. And these hens had to make their journey quickly, with bodies still light and agile enough to leap up the fish ladders and rapids. If their bellies were too distended with eggs, they wouldn't manage. If and when they reached the mouth of Eagle Creek, they would fast for months as their eggs reached maturity while the bucks, fierce with sexual energy, fought to be their mates.

When the warm rains of early spring came, the hens gathered strength for the last time, their bellies full and ripe and ready to give life. Their last ordeal was to make their way up the gentle rapids of Eagle Creek at the optimal time. They didn't want to go before their eggs were ready, but they didn't want to go too late either. Too late meant a dried-up creek and a zero return rate of their offspring.

The steelhead of Eagle Creek were unlike any other steelhead Guido knew. In the river, he could recognize them instantly. When he hooked them, they went berserk, running fast and far downstream, forcing him to chase them on foot through the water. They were strong and stubborn and they resisted with every ounce of strength they had. When he managed to reel them in, he always took a good look before releasing them. They were shaped like bullets, deep-shouldered and broad across the back. Their sides were rust colored and their backs green gray. They weren't the biggest steelhead out there, but they were among the wildest. Guido loved their strength and spirit, and he rued the fact that he saw them less frequently.

As these wild fish disappeared, something else was happening. We began to see more hatchery fish. As salmon and steelhead vanished from the Columbia Basin, fisheries managers were doing everything they could to make up the numbers by producing hatchery fish. There were hatcheries popping up on rivers everywhere: the Yakima, Grand Ronde, Snake—most of the major rivers that fed into the Columbia now had hatcheries in furious production. In the early nineties, Oregon's fisheries managers were taking another precautionary measure. They were transporting the baby fish by barge and releasing them below the deadly turbines of the Bonneville Dam. Which presented a problem. For fish birthed from a barge, where was home? For many of them, it became the Deschutes River. The reason had to do with the temperature of the water.

When the steelhead and salmon of the Columbia Basin returned from the cold Pacific, they faced the shock of an increasingly warm Columbia, which by late summer was flowing at an uncomfortable, oxygen-thin 70 degrees. Branching off this mother river were their cold natal rivers. The fish had the smell of these rivers locked into their inner compasses and doggedly made their way as quickly as possible back to the source. But first they had to face the monumental obstacle of the Bonneville Dam, a soaring wall of concrete that spanned the Columbia 130 miles from its mouth. Off to the side of this churning maelstrom of water was a fish ladder. It was the biggest challenge of their journey, but before they even reached it, migrating fish had to survive the stress of the river's warm water as they darted from cold pocket to cold pocket, navigating gill nets, feasting sea lions, birds, and thousands of fishermen. Only some of them made it.

Like the others, the homeless hatchery fish exhausted themselves jumping up level after level of this fish ladder, fighting the rush of the artificial rapids one after another. Once they made it through the gauntlet of the river's largest dams, the Bonneville and the Dalles, the first delicious cold-water plumes to reach them came from the Deschutes. Wild fish belonging to other rivers swam on, but hatchery fish slowed at the smell of this cold, oxygen-rich water. Many of them decided to exit then and there, taking a hard right away from the balmy Columbia into the clear, spring-fed waters of the Deschutes.

By 1995 the population of hatchery fish in the Deschutes had sky-rocketed. These were now the only fish we could legally keep. The wild stock was too small, and too precious. One season when we were hiking up Eagle Creek, we watched a wild fish mating with a hatchery fish. In the clear shallows, it was easy to see the clipped adipose fin of the hatchery fish. Guido was unsettled by it. He was one of a growing number of fishermen who recognized that hatcheries were accelerating the demise of their natural-born brethren.

In the beginning, hatcheries seemed like a logical solution: if you didn't have enough fish, just make more. The problem was that people couldn't come close to mimicking the natural process. Hatcheries used only a few fish to create millions of offspring. Imagine taking three humans, selected for their strength and general health, and cloning them five hundred thousand times each, and setting them loose in any city on earth. The moment these clones entered the population, they would begin competing for the same resources as the natives—vying for the same homes, schools, partners, and jobs. As the clones interbred with each other and the natives, there would be an inevitable dilution of DNA (which is why it's ill-advised for close relations to procreate; too much genetic similarity creates anomalies and maladaptations in the offspring). In the salmon of the Columbia Basin, and all up and down the Pacific Northwest, the same genetic combinations paired over and over again weakened the general population.

Scientists call the phenomenon "genetic drift." Salmon were suffering from the very loss of genetic diversity that had once assured their abundance. What remained were a lot of fish that were extremely similar, which would have been fine in a world that was totally stable, without any fluctuations in the oceans and rivers. In such a world, it was possible that this homogenous population could survive. But in the real world there were variations—some years brought extreme weather, drought, and unseasonable temperatures, all of which challenged hatchery fish beyond their genetic capability. In these years, the results were catastrophic; of the millions of fish that had entered the Pacific, few returned. For however strong and healthy the clones were, they diminished what was emerging as the single most important component to a healthy salmon ecosystem—diversity.

In the case of Eagle Creek steelhead, when the hatchery fish overwhelmed the wild population, the fish no longer knew exactly what to do and when to do it. Over the years, as Guido described the situation, it was like notches in the key got worn away. Soon the key no longer turned the lock. In time, the native messages of survival became fainter and fainter and eventually were lost.

The more I thought about the term "wild fish," the less sense it made. It seemed there was little that was truly wild about them. They lived with a high level of reason and order, perhaps higher than we humans can grasp. I have thought about how many creeks and streams there have been like ours, with fish that have been formed by their unique water flow and temperature and season, by all the micro-conditions they must survive in order to reproduce, and how their wildness is a word we use for something we can't completely know. We can't understand "wild" nature, because we haven't been here long enough to understand it. The wild steelhead of Eagle Creek had behaviors that had developed over tens of thousands of years, which made them part of a distinct history that belonged to an exact place, that once fit as neatly as a lock and key.

Hatcheries were now under scrutiny in Oregon, where the state was still embroiled in the aftermath of the Endangered Species Act, and Guido was at the center of the storm. He had taken a job with Oregon Trout, the feisty little fish conservation group that had made the ESA petition for sockeye. By now Guido was a seasoned fund-raiser, and he understood how to build organizations. He'd been hired to help grow Oregon Trout into a legitimate, directed outfit, and they had given him the green light to advance his stronghold platform. He'd been busy writing letters and traveling the state, deliv-ering arguments about protecting the healthy rivers they still had. He'd identified a flaw in the system that few seemed to recognize. The Endangered Species Act was not a conservation strategy—it was the emergency room. By the time a species was endangered, the whole system was failing. It was code blue; life support could be adminis-tered at great cost, but a full return to health was out of the question. Preserving strongholds was such an easy way to prevent these rivers from following the crippling path of dams, clear-cutting, and hatcher-ies. Dams blocked circulation, clear-cutting raised the temperature, and hatcheries were like an infusion of tainted blood. It was a death sentence, and it could be completely avoided. His efforts came to nothing. Stepping back, Guido understood why his message was fall-ing on deaf ears. The concept of prevention was too much for an overloaded system still in shock over a fatal diagnosis. Oregon re-mained stymied by the fundamental question: What had happened to their salmon? Until this question was answered to everyone's satisfac-tion, there would be little room for new ideas.

The federal government had now entered the scene. Oregon was proudly independent, and it found itself unhappily entangled with an intrusive and cumbersome federal bureaucracy that struggled to respond to an environmental crisis it didn't fully understand.

It wasn't for lack of trying. In an attempt to establish a base level of clarity on the new "endangered" designation, the US National Marine Fisheries Service had raised more questions. They were essential questions, and they needed answering before any plan could be put into place. First off, what constituted endangered? For that matter, what constituted a species? When it came to Pacific salmon, the ESA offered no initial guidance for this determination. They came to agree that a species of salmon represented a "distinct population." "Distinct" for ESA purposes meant that it represented an evolutionarily significant unit, or ESU—a mystifying status awarded to 1) a population that was reproductively isolated from other populations and that 2) represented an important component of the evolutionary legacy of the species.

If a species passed these two tests, its protection would be guided by the "best scientific information available." It seemed to make sense, but Pacific salmon posed a few inherent problems. If an endangered species of salmon was by definition unique and isolated, it meant that one could not study one population of salmon and assume the information was true for another. The salmon of the Deschutes River had completely different requirements from the salmon of other rivers. How, then, to make a comprehensive plan? How to draw up anything in black and white with this multicolored ecosystem? Underlying these questions was the biggest dilemma of all: How could one establish a healthy standard by studying a sick population?

The issue of determining a species's "evolutionary legacy" led to perhaps the most unanswerable questions: What was lost when a species became extinct? Did the genus suffer from this loss of diversity? Was diversity important, and if so, why? Somehow the job of defining the nature of nature had fallen to a governmental bureaucracy, a machine made up of myriad gears that did not always mesh; rather than rolling the process forward, it often led it deep into the weeds.

As far as Oregon was concerned, the worst possible outcome was to have the federal government mucking around in sensitive state issues. The spotted owl had been hard enough. No one could have predicted the fallout of this owl being listed as endangered. It had forced the strangest contest ever: a medium-sized forest-dwelling bird versus the oldest and most lucrative industry in the state. Incredibly and infuriatingly for many, the owl won, and the practice of clear-cutting the old-growth forests was brought to a halt. Logging on federal land was dramatically decreased, and new rules were instituted for logging on private land. This had meant lost jobs and business and state income.

What then, would happen with salmon? Salmon were everywhere, from the ocean near Oregon's long coastline to rivers and lakes hundreds of miles inland, where the fish threaded their way through a wide range of business concerns: commercial and recreational fishing, property development, large and small agriculture, hydroelectric dams, forestry. These were huge industries that would all be opened to federal scrutiny, intervention, and new legislation.

In 1992 it was still unclear how intrusive the federal involvement would be, but the state was smarting over the owls. When, a year later, Oregon Trout prepared to petition the federal government to classify coho salmon as endangered, the state pushed back. Coho spawned in rivers up and down Oregon's coast. If they were listed as endangered, the whole coast would be opened to regulation. This was when the governor of Oregon stepped in and gave the little organization a call.

Governor John Kitzhaber was an unusual public servant. He loved Oregon's wilderness, and knew its fish and rivers well. An outdoorsman and an emergency-room doctor, he took an approach to problem solving that was unfettered by bureaucratic considerations. He was distinctly apolitical and had a knack for reaching across party lines to achieve partisan consensus. Notorious for his cowboy boots and jeans, he played things straight and sought long-term solutions. What he saw with the pending coho salmon petition was a worsening of a conflict that involved the state's most productive industries. Foresters, commercial fishermen, developers, and farmers were being held accountable by conservationists who were calling attention to the fact that there was a crisis. It was a serious problem, but for Kitzhaber it was Oregon's problem.

Kitzhaber contacted Oregon Trout and proposed a solution that wouldn't involve the federal government. He offered Oregon Trout a deal: if they withdrew their petition, he would negotiate with the state legislature directly to take the measures necessary to help the coho recover. His case was that he could win more conservation measures this way. His leverage with the state would be the raised hammer of another ESA petition.

The strategy worked. Oregon's spirited tradition of independence and its aversion to the feds kept the hammer from falling. As long as the state worked toward conservation, another ESA listing wasn't required. Over the next year, the legislature agreed to implement Kitzhaber's requests to reduce both hatchery production and overfishing. It was called the Oregon Plan for Salmon and Watersheds, and Kitzhaber met with his agency heads every week to check on its progress.

For Kitzhaber it was only a partial solution. Awareness had to be effected on the ground—with the people. This, he believed, was the only way to achieve lasting change. He decided to create watershed councils for Oregon's rivers. These councils were made up of stakeholders, or those whose lives and/or livelihood involved salmon habitat. Each river zone had its own dedicated group of local fishermen, farmers, loggers, and conservationists who were instructed to sit down and talk. The councils were advised to listen well to one another, for they had been tasked with overseeing the health of the rivers as a team.

Kitzhaber maintained that this collaborative, grassroots action was the heart and soul of conservation, and he was proven right. The watershed councils talked, and listened, and grew to appreciate one another's positions. In time, they were able to reach consensus on some pertinent issues, agreeing on measures that would both protect salmon habitat and allow industries to thrive. While Oregon's coho were eventually listed as endangered, it was through negotiation and peacemaking that they became one of the first endangered species to return from the brink and later be listed as "likely to recover."

Guido watched the whole process with a growing admiration for Kitzhaber. The key, he saw, was communication and incentive, not lawsuits. He also saw how the threat of action could be more effective than action itself—a raised hammer was more powerful than a hammer wielded. Kitzhaber's principles and his integrity made a deep impression on Guido, and in the coming years he would learn much from the older man.

While he was getting nowhere with strongholds, Guido persisted on his own, marking the healthy rivers throughout the Pacific Northwest for protection in the hopes that someday people would be ready for such a measure. In the meantime, he had succeeded in bringing Oregon Trout out of its infancy, raising money to hire a professional director and recruiting new board members who carried a weight that would register on the national scale. All the while he kept his mind on strongholds and the majestic peninsula that, on his new map, lay just across the water. As he wrote papers and grant proposals detailing his stronghold mission, the beauty, sense, and feasibility of his plan gestated inside him: protect the places that were still whole, before they too were lost and all memory of wholeness lost with them.

He was becoming ever more aware that his role was to coax the right people into the river with him. There, with the wind in the trees and the gentle tumble of water over rocks, they could experience the exhilarating world of a salmon river. With some artistry, luck, and a fly rod, they could have a direct connection to the electrifying force of a migrating salmon. Some might glimpse a realm they had never seen before. It could change a person; it had changed Guido. He could tell the story, but the river and the fish could tell it better. He just had to make the introduction.

CHAPTER 9

CATCHING LEE

At thirty-three, Guido was becoming increasingly aware of his own biological clock. He'd had a few relationships since college, but it had been years since he'd met someone who held his interest. He'd been keeping his eye out, going to parties and doing his best to circulate, but he hadn't felt that "click." While his professional life moved forward, he felt stymied by his extended bachelorhood. His solitude, once so fiercely defended, was something that now seemed negotiable. The trouble was, he told himself, he hadn't met the right person, though part of him wondered if such a person existed. More likely, there was something wrong with him.

In the winter of 1993, Guido went home to Minnesota, where there wasn't much to do in the snow-covered landscape except socialize with his family and old friends and tool around with his younger brother Willie. On his last night there, Willie suggested they visit his friend Lee Lane. Willie drove them down the icy country roads that led to a cheerful little shop called Frost and Budd where Lee was working. Guido was half awake, staring out at the snow-muffled landscape that harbored nothing of interest, not even in hibernation.

After pulling up to the quaint gift shop, Willie and Guido kicked their boots free of snow and shuffled in. It was a sweet and happy place, filled with candles, Christmas ornaments, and other tchotchkes, somewhere Guido would never have found his way to on his own. As the Christmas lights winked and the holiday music drifted over the speakers, Guido's hunter's gaze relaxed. He was thinking of other things and places he'd rather be when he raised his eyes to meet Willie's friend.

Lee was a beautiful, outdoorsy Minnesota girl with unusual poise and charm. By the time Guido came on the scene, she had broken more than a few hearts. Warm and direct, she gave the brother of her friend a hug. She knew of Guido because on the employee wall of Frost and Budd was a letter he had sent them on letterhead decorated with a hand-drawn humpback whale. The few polite lines suggested that the shop look into recycled materials for their shipments. Guido had received a gift from them, a small vase bought by his mother, and it had arrived encased in a veritable sea of styrofoam. This, he said, was an unnecessary waste.

When Lee embraced him, Guido felt as if he'd been electrocuted. He stood speechless as Lee chatted. In the cluttered shop, Guido heard every third word, and for some reason, he was taking steps backward, as if needing to view this woman from a distance. He stepped back far enough to upset a potted fern, momentarily snapping out of his trance to prevent it from toppling.

The two brothers lingered until Lee was called back to work. Willie watched his brother stagger out of Frost and Budd and sit mute in the passenger seat of the car. As they drove home, he told Guido to simmer down; Lee was not available, and for a number of reasons. Guido listened without hearing. His bell had been rung and he was silently vibrating, thrill and terror rippling through him in equal parts. He was distantly registering Willie's words: Lee had a boyfriend. Lee was joining the Peace Corps in three months. It was not to be. But Guido had been told no before. "No" was not definitive; it simply meant finding another way. With Lee, Guido had no choice. The hook was in his mouth, and there was nothing to do but fight it out until something released him. For now, he needed to know more. He needed to see her again.

He didn't have to wait long. When he discovered that Lee, Willie, and some other friends were heading to Oregon to go fishing in a week, Guido attached himself to the trip, suggesting to Willie that in addition to going fly-fishing, they should head to the coast to the Necanicum River, where he could teach them how to steelhead. If it was Guido's instinct to show himself to this potential mate in his natural state, it was a good one. Lee would soon see exactly who he was. Then she could make up her mind.

The Necanicum was a small river that flowed through a dark, temperate rain forest of Douglas fir, alder, and cedar. It was rainy, and the river was high, and the trees dripped in the mossy forest. The men in the group had some experience with steelheading, so Guido cheerfully offered to instruct the women. He explained that steelhead fed on the bottom of the river instead of rising to the insects that floated on the top, like trout. Though much bigger, they were harder to detect.

Lee listened rapt as Guido described how the river before them provided "lies," or natural places of rest where the steelhead would shelter to gather their strength. This is where they would direct their flies, which sank below the surface so they could reach the depths where the steelhead were holding. Guido then waded into the river and demonstrated some casts, carefully probing pockets and runs. Lee watched him standing hip-deep in the water and found it hard to look away; she had never seen a man so at ease, so in control in the midst of this mysterious world he somehow owned. And she was totally game. In Guido's borrowed waders, she stepped into the river with a smile on her face. As the rain drove down from the sky, Guido told her to keep her expectations low; the river was blown, and catching fish would be hard. Lee nodded, already absorbed in casting her heavy, wet fly perfectly into the murky water.

As the days passed, Guido found himself talking to Lee about things he had never shared with anyone. Finding ways to be by themselves, they laughed giddily at silly things. To the rest of the group, their infatuation was clear. In Guido's opinion, the fishing trip ended all too soon. As they drove back to Portland, Guido suggested that he and Willie take Lee out to the Deschutes to see what the fish were biting on there. For Lee, it was an easy yes. Guido drove them in his Volkswagen Vanagon up the Columbia Gorge, with its towering monoliths of rock, with Lee by his side. He told her about the Columbia River and that, of the three rivers that cut through the Cascades to the Pacific, the Columbia was the mightiest, and that salmon had once filled its waters.

Later, when they'd gotten settled at the Deschutes, Guido and Lee went for a walk. It was a gorgeous evening and the high desert and canyon land unfolded around them. Lee was struck by the dramatic landscape, and could sense how important this place was to Guido. Overwhelmed by her attraction to him, she tentatively started a conversation. "I don't know if you're feeling what I'm feeling . . . ," she said, letting the sentence hang.

"Oh, I'm feeling it," Guido assured her. "I'm definitely feeling it."

The brothers gave Lee the master bedroom. When Guido turned down the covers for her, he pulled up a chair and opened a book he'd dug out of the cabin's little library. He read from Wallace Stegner's *Beyond the Hundredth Meridian,* chosen perhaps for its recounting of the life of geologist and ethnologist John Wesley Powell, a bold and eccentric man who foresaw the environmental demise of the American West but could make no one listen.

The next morning, the three of them headed downriver with fly rods. The day was balmy and mild and the hours passed quickly. For lunch Guido laid out cheese and crackers, and they drank a bottle of Chianti while he regaled them with stories of other rivers he had known, watching Lee closely all the while. This was, he realized, the hunt to end all hunts, and he had limited time to make a lasting impression. As they sat together on the smooth rocks above the water, he noticed that Lee was getting the hang of casting. He also noted her comfort on the Deschutes, and something on her face that looked like happiness.

Two days later, Lee flew back to Minnesota, sharing with Guido that in a few months she would be joining the Peace Corps, where she'd been assigned to sustainable farming somewhere in Central America. As lovely as it had been to spend time with Guido, a relationship wasn't going to work. She would be gone for two years. They were going to miss their chance. It happened. There was nothing to do but say goodbye.

Guido felt a sharp and unfamiliar pain and realized he was in love. While they had no future, he felt compelled to tell Lee how he felt, and how much it meant to him to have met her. He had been waiting to feel this kind of love for years. Until now, he'd thought there was something wrong with him, that somehow he existed in a place where love couldn't find him. But he knew now what it was supposed to feel like. He thanked her because she had shown him, as quirky as he was, that love was possible for him. Then he gave her a gift. "Open it later," he said.

At home, she sat on her bed and unwrapped Guido's gift. It was a copy of *The River Why,* David James Duncan's coming-of-age novel about an obsessive young fisherman who one day finds love. Slipped between the pages was a note. "Lee, I will never forget our days on the river together, and hope there will be more. Love, Guido." Tucked in with the envelope were two sprigs of sagebrush. If Guido had a scent, it was this pungent, purifying smell of the high desert.

She called him that night. It was still early in Oregon, and after a moment of small talk, they both knew it was time to acknowledge that something serious was happening between them. Guido had a few months to beat the calling of the Peace Corps. For the next week he ate and slept little. Every few days he wrote Lee a letter, telling her how much he missed her. He enclosed fresh sprigs of sagebrush and the draft of an article he had written for *Flyfishing the West* magazine about his trip to the Bolshaya with Misha Skopets. In the following weeks, Lee received more manila envelopes stuffed with sagebrush, drawings of fish, and even VHS tapes of his *On the Fly* program. Guido tied every fly he had and cast it toward Lee, and then he waited.

Lee broke up with her boyfriend, telling him that her heart had been stolen by someone else. She then went through the motions of her day while the wainscoting ledge around her room filled with Guido's sage. She missed everything about him. That March, in the dead of Minnesota's lingering winter, she surrendered, packed up her white Subaru, and headed west, taking a job with the Xerces Society, an international invertebrate organization that worked primarily with butterflies, a species as fragile as salmon were tough. Guido and Lee moved in together that Fourth of July. In the coming months, their courtship unfolded at the Deschutes, where Guido inducted Lee into his world. In the evenings, they would take gin and tonics down to the river, and Guido would teach her how to improve her cast, patiently untangling her often snarled line.

Guido moved things forward as he always did, as if the clock were ticking. Timing was as critical in romance as it was in the natural world, and he did not want to make the mistake of waiting too long. It was December, only a year after they'd met, and they were at the Deschutes when Guido made the odd suggestion that they go for a hike. Usually he preferred to stay on the river, but today he seemed game for a walk up the shoulder of Mutton Mountain. Their conversation was light on the way up to the ridge, but when they stopped, the tone changed as Guido told Lee how much he loved her, and how happy he'd been since they'd met. Without pausing, he declared his desire to raise a family with her and watch their kids grow up on the Deschutes. Suddenly, in front of her in the cheatgrass, he dropped to one knee. Lee stared in amazement, totally unprepared. There were so many unknowns. Guido was charming, but he was also *different*. What would life with such a person be like? She would have no time to answer such questions because, while her pause couldn't have been more than twenty seconds, Guido's face was showing tension, if not panic.

Then the pause ended and Lee smiled and said, "Of course I'll marry you."

High above the river, as the sun sank below the hills, they drank a bottle of Dom Pérignon. Afterward, they stumbled down the ridge in the dark, trembling at what they'd just done.

They could have waited a year, but they couldn't find a reason to.

That summer, Guido married Lee Lane at the Rahr ranch in Montana, on a thousand-foot cliff overlooking the confluence of the Smith River and Tenderfoot Creek.

Guido had told Lee about strongholds, and what he wanted to do with his life. He talked about Kamchatka and how he needed to find a way back because it was possibly the biggest salmon stronghold on earth. Lee both understood and believed in his mission. Her support gave Guido an emotional stability he had never known. In years to come, she would provide a base that steadied him as he flew high and far.

CHAPTER 10

COLD WAR WARRIORS

BY 1998 GUIDO and I were both married and settled in Portland, and once again leading parallel lives. Lee and I became
friends, and our friendship grew strong at the Deschutes. We carried
on the traditions of our parents, hiking the mountains and canyons
together, speaking about the world and our lives in the place they
were the clearest. The ancient landscape, with its rock canyons,
craggy escarpments, and majestic river lent us a dignity, just by being
part of it. Skirting the escarpment high above our cabins, we found it
easier here to accept realities that we resisted back home: that our
time on earth was limited, that we were small in the scheme of things,
and that the many dramas that absorbed our day-to-day existence
didn't matter much in the end. Our gratitude for the place that returned us to these truths lightened our earthly loads and somehow
made the complicated job of being human easier.

My friendship with Guido had deepened. Sometimes I was reminded that our parents had been siblings, and that we shared a good
amount of DNA. A first cousin is a more distant mirror than a sib-
ling, but can reflect certain similarities. As we grew older, Guido and

I seemed to grow more alike—or maybe we just understood each other better. I still trailed after him sometimes when he went fishing, and we fell into an old and comfortable silence walking along the river while he scouted the water.

That year Guido found a way back to Kamchatka, partnering with a man who was unlike anyone he'd known. Ex–Navy Captain Pete Soverel had commanded major ships of war and served two presidents, the defense department, and NATO. He had won medals for bravery in two wars and taught a course at the University of Washington called, simply, War. He was also obsessed with fishing for steelhead. It was an obsession that changed his life, and ultimately led him to Guido.

Pete told Guido how he had returned from the Cold War to find his home rivers in the Pacific Northwest depleted of fish. This new reality had sent him on a rampaging inquiry—what had happened to the steelhead? Who in the chain of command was accountable? He read articles and contacted agencies and experts in search of information about how such an important fish could simply disappear. When he found no satisfactory answers, Pete initiated action of his own, fighting against the building of new dams and petitioning to get struggling steelhead and salmon populations listed as endangered. When it seemed to Pete and his small group of angler friends that no one was advocating for the protection of steelhead, he and his friend Tom Pero, longtime editor of *Trout Magazine,* founded the Wild Salmon Center. It seemed steelhead protection would come down to citizens.

Pete, who abhorred the red tape of bureaucracies, had no idea how hellacious a process protecting steelhead would be, especially when it came to listing them as endangered. Of all the salmonids, steelhead were the trickiest to describe because so little was known about them. One could not list a species as endangered without proposing a plan to protect them. Here Pete's frustration doubled. It seemed there wasn't enough data on steelhead to protect them—how did one protect something one didn't understand? In Pete's militaryspeak, what was needed was more intelligence on the species, intelligence that, due to gross mismanagement, was impossible to gather on degraded American rivers.

But Pete possessed unusual resources when it came to intelligence. His good fishing buddy John Sager was another fanatic steelhead fisherman—and he was also a trained intelligence officer with a long and illustrious career working for the CIA. Sager had spent most of his time in the agency's Soviet division, serving as the acting Moscow bureau chief during the Cold War as well as a top-level spy. Sager was just as interested as Pete in taking action to protect steelhead stocks in the Pacific Northwest, and was equally alarmed that so few cared about the disappearance of a fish that had once numbered in the hundreds of millions. The two men fished together in the rivers outside Seattle and strategized. They decided to start a newsletter dedicated to steelhead, which other fishermen instantly rallied around, including Guido. It turned out that they weren't the only disgruntled anglers out there. *The Osprey* soon became a leading authority on steelhead in America, publishing well-respected articles that were submitted from a variety of sources, from fishermen to scientists. The little publication went some way toward shedding light on their elusive subject, but it did not bring the fish back to the rivers.

Then, in 1992, Sager stumbled upon the most riveting "fish intelligence" he had ever seen. Through his old Soviet network he was passed an academic paper translated from Russian. Written by a Russian female ichthyologist named Ksenya Savvaitova and published two decades earlier, "The Noble Trouts of Kamchatka" showed that legitimate research had indeed been conducted on steelhead, and in Russia no less. Cold War warriors like Sager and Pete knew about the military-only zone of Kamchatka. This was the highly protected peninsula where the Soviet nuclear fleet was based, and access was strictly prohibited. Geography suggested that this was prime salmon territory; the question Pete had long entertained was, Were there also steelhead? "The Noble Trouts" answered this question definitively, and with remarkable findings that had lain hidden from the West for the duration of the Cold War.

Published in 1973, "The Noble Trouts of Kamchatka" was possibly the most comprehensive study of steelhead ever conducted. Savvaitova had been traveling to Kamchatka since the late fifties, staying months at a time to observe the unusual fish. She had been tasked with shedding whatever light she could on the enigmatic *Oncorhynchus mykiss*—a fish that was both trout and salmon, with anomalous behavior that could be explained by no one. It seemed the Russians were just as mystified by steelhead as the Americans. In the coming years, Savvaitova set to work on establishing the differences between the resident and the migratory form of rainbow trout, or steelhead. By the early seventies, she had visited "her" Kamchatka steelhead more than twenty times, staying for periods of up to three months and gathering a veritable trove of information. It took years for her to condense her research for publication—and two decades for her paper to reach an intensely eager audience on the other side of the world.

After skimming the content, Sager got the paper to Pete, who tore through it, immediately grasping its implications. Kamchatka was full of steelhead, and they were a remarkable population. But even more, here were decades of research conducted by someone who had observed steelhead in a pristine, undisturbed environment, living as they had lived for millennia. Savvaitova was like a time traveler who had gone back two hundred years and reported on rivers as they might have once been. Kamchatka's isolation had perfectly preserved its fish and their habitat, and for those who were losing their salmon and steelhead rivers, this presented an opportunity. If Westerners could learn more about steelhead in the uncorrupted rivers of the Russian Far East, they might understand how to protect them back home.

Pete started gunning for Kamchatka. Not only did he want to explore its rivers with a fly rod, he wanted to know if the rivers were as intact as Savvaitova had left them, and if they were still full of steelhead. He quickly found out that while gaining access to the once-forbidden peninsula was difficult, fishing for steelhead there was impossible. Before she'd closed down her research efforts in the

1970s, Savvaitova had listed Russian steelhead as endangered. Fishing for them was illegal.

It was with the help of some of his Cold War buddies that Pete finessed a deal with the Russians. It was a simple exchange, Pete explained to Guido: science for fishing. Russia was extremely interested in steelhead, but it had been decades since there had been funding for such projects. They of course would welcome free research, if this was what Pete was offering. It was. Fly fishermen (who did not exist in Russia) could double as researchers, Pete proposed. It sounded simple enough, but, as Guido would learn, nothing in Russia was simple.

As luck would have it, Pete's Cold War network reached all the way to Moscow, to Sager's old friend Serge Karpovich, another CIA Soviet division veteran. A Russian-American, Karpovich had ascended to the highest rung of the CIA's counterintelligence division during the Cold War, interrogating famous double agents and handling notable defectors such as Alexander Solzhenitsyn. After retirement, Karpovich set up an intelligence and security agency in Moscow. He had cleverly partnered with his cousin Gennady Inozemtzev, an intelligent and connected man who had been in charge of reverse engineering for Russia. Karpovich found a third partner in his old enemy, the now unemployed General Viktor Budanov, once known as the most dangerous man in the KGB. A one-time boss of Vladimir Putin, Budanov had been in charge of Directorate K, a division within the KGB responsible for internal security. Directorate K essentially spied on the spies. It was referred to by some as SMERSH, a Russian portmanteau meaning "death to spies." The combined intelligence of the three men was copious, and amounted to a highly select database perfectly positioned to help foreigners navigate Russia.

In 1993, John Sager presented his friend Karpovich with the strangest assignment to date. A newly minted two-man organization that called itself the Wild Salmon Center wanted to find a way to fish for steelhead in Kamchatka. Karpovich, an amateur angler—Sager had taught him how to fish when they were stationed in Iran together—accepted the job. His initial inquiries revealed that the key to accessing steelhead in Kamchatka lay with Moscow State University, where

practically every living thing was studied and documented. Karpovich explained to his Western friend that in Russia, significant species
had designated scientists, or "godfathers," who dedicated their lives
to studying a single classification. These scientists were the gatekeep-
ers of their species. In the case of steelhead, it turned out that instead
of a godfather, there was a godmother, who was none other than
Ksenya Savvaitova, author of "The Noble Trouts of Kamchatka."

By 1994, Savvaitova was in her seventies and head of the Laboratory of Systematics and Fish Ecology at Moscow State University.
Karpovich met her at her office at the research wing of MSU, which
had been gutted of its funding, and much of its electricity. Savvaitova
was alone in the dimly lit building save for a single research assistant.
She was round faced with pretty features partially obscured by thick,
oversized glasses. While highly skeptical, Savvaitova was both hum-
ble and kind, and she listened to Karpovich politely as he made his
request.

She was instantly confused by this group of Westerners who wanted
access to steelhead in Kamchatka. Why would they want to come so
far for fish? Anyway, steelhead were an endangered species—she had
listed them in Russia's endangered-species Red Data Book herself. No
one should be fishing for them! She added that since funding had
been cut off, she hadn't been able to visit her Kamchatka steelhead,
and she was worried. Poaching was a terrific problem. Steelhead from
her rivers were being served in Moscow's restaurants.

Karpovich explained that the men he was representing were fly
fishermen and weren't interested in killing the fish; they were "catch-
and-release" fishermen. Savvaitova had never heard of such a practice. The bottom line, Karpovich told her, was that these anglers had
no interest in reducing the steelhead population. In fact, the Western
fishermen might be able to both underwrite an ongoing research project for Moscow State University and aid her with the scientific research. Savvaitova could instruct them on how to measure, weigh,
and scale their catch, adding substantially to her sampling of the
population. They could work together. It was a win-win proposal.

At the prospect of resuming her work in Kamchatka, Ksenya Savvaitova overcame her resistance and started the laborious process of obtaining the permits that would allow Pete's group to legally catch steelhead in this top-secret military zone. Because of the scientific integrity of her proposal, she was eventually granted her permits, and with them came the keys to the kingdom of Kamchatka. In 1994, Serge, Ksenya, and Russia's most esteemed ichthyologist, Dmitri Pavlov, joined as founders of the Wild Salmon Center.

Pete confirmed to Guido that working in the Russian Far East was hard, not to mention expensive. The logistics were beyond challenging, and issues with equipment, weather, and alcohol consumption among the Russian crew had worn Pete out. He was tired, short on money, and ready for help. Guido Rahr, elite angler and rainmaker, had already fished in Kamchatka and would have an idea of what to expect. Would he be interested in a job?

Guido was extremely interested. Pete drove down from his home in Seattle, and the two met for lunch in downtown Portland. The contrasts between them were undeniable. Pete was tall and fair, with sharp blue eyes and unquestionable command. Guido was lighter, quicker, with a disarming openness, and he had never taken orders from anyone. Over lunch, Pete sat back and described how he had partnered with Russia's two top ichthyologists to create the Wild Salmon Center. They ran two research-fishing trips a year to Kamchatka, presiding over a successful joint project that gave the Russian scientists the data they needed and the American anglers the fishing experience they craved.

When Guido accepted the job as executive director of the tiny organization, he had little idea of what he was getting into. Nor did Pete know of Guido's agenda, or what the younger man was capable of. Guido took a week to study Pete's operation and saw an opportunity. Then he set to work on a memo outlining what he sought for himself and for the Wild Salmon Center. It was part vision statement, part manifesto. Guido would do as Pete had asked, and take over his steelhead project, but he wanted something in return. He suggested they pull out from the map a few steps, and a few steps more, until the bigger picture came into focus. He told Pete he'd join him on the condition that they focused not just on the steelhead of a few rivers in Russia but on all the salmon of the Pacific Rim.

Guido's rat-a-tat bullet points described the stronghold vision, and how critical it was to save the best 15 percent of the earth's habitat, which in this case meant the best salmon rivers of the Pacific Rim, from California to Japan. It would require the diplomacy, research, and compliance of many nations, first and foremost Russia. It was a monumental undertaking, and it would be fraught with challenges.

Guido knew he was taking a chance. Pete was looking to lighten his load, not increase it exponentially. The memo represented a huge mission leap, and he had no idea how Pete would react to it. Maybe Pete was tired of being alone at the helm, or maybe he was just ready to set a different course, but Guido's memo had an unexpected galvanizing effect on the older man. The stated goal was absurdly ambitious, but Pete liked it, and he liked Guido's gumption. When they met to discuss it, Guido was surprised to hear Pete say, "Let's do it!" adding the qualifying jibe, "I'd rather be shot on the advance than the retreat."

In the fall of 1998, I went over to Guido's house for dinner. He and Lee had bought a house in the Southwest Hills neighborhood of Portland that looked out over the city, Mount Hood, and the Willamette Valley. On clear days the coastal range was just visible in the distance. Guido was leaving for Kamchatka the next day. He would be spending three months at Pete's camp on the Utkholok River, where he would learn the ropes of the Kamchatka Steelhead Project. Guido paced around the house, talking excitedly about the Wild Salmon Center and how things were falling into place for him. But there was a tension in him, and in the few glances he and Lee exchanged.

During dinner, Guido kept leaping up and pacing, as if trying to remember something he had forgotten. I don't think I'd ever seen him so on edge. Finally he explained that the sealed box in the hall was filled with fourteen radio transmitters. Oregon's Department of Fish and Wildlife was sending them to the Russian scientists of the Wild

Salmon Center to aid in their steelhead research, and Guido was in charge of bringing the box to Kamchatka. To ease his passage through Russia, he had been instructed to strap $75,000 to his body. It was a cash economy, Pete had told him, and paying people with fresh green dollars was part of the system.

For all their intelligence and capability, Pete and Karpovich's team of spies cast a long shadow for their fledgling organization. They had set up shop in the backyard of their defensive, paranoid, onetime enemy, and it hamstrung them from the start. In a country where trust was a maladaptation, the Russians found ongoing reasons to suspect Pete and his crew. Kamchatka had secrets it was not willing to part with, and the names of old enemies were not easily forgotten. Within Russia's internal security apparatus, there were still red marks next to the names Soverel, Karpovich, and Sager. Even after the collapse, security was security, and Russia's enemies had not changed, especially the one that lay directly across the Pacific from Kamchatka. Guido Rahr's name was likely added to the Federal Security Service's (FSB) watch list the day he arrived, and the transmitters were confiscated at customs as they were no doubt some Western spy equipment.

Who could blame them? Misha later asked. What was Russia to think about an organization that employed former high-level KGB and CIA operatives and brought untold amounts of money to the Russian Far East—the most top-secret region in all of Russia—purportedly to fly-fish?

Pete's camp was in western Kamchatka, on the banks of the meandering, tea-colored Utkholok River. In the middle of a vast tundra plain sat a cluster of yurts that housed the small group. Marking the vulnerable little settlement were the American and Russian flags, flapping in the wind on separate poles. There was nothing else for as far as the eye could see. Guido would spend the next three months in this remote land with its arctic foxes, bears, and skies filled with autumnal bird migrations. Occasionally Koryak horsemen passed by with their rough saddles and bone stirrups; otherwise there was no one.

Russian scientists and American fishermen gathered for meals but spent their days apart, with a handful of anglers fishing for steelhead up and down the river, and the two Russian scientists analyzing their catch in their research tents. A Russian staff kept the camp running, with no small amount of grumbling over the fact that Pete had limited their alcohol consumption to one bottle of vodka at meals; there had been too many accidents with equipment. Pete had employed a few American guides as well, elite anglers who acted as Pete's "captains," and had authority over the Russian staff. Guido noticed that it made for some tension between the two groups.

Ksenya was one-half of the scientific team from Moscow State University. The other half was the esteemed Dmitri Pavlov, whom Russians called *bol'shaya shishka,* or "big cheese." It had been Dmitri's clout that had helped sway Moscow's decision makers to sanction the unusual arrangement with the American group. Ksenya spent most of her time in her research tent, processing the fishermen's data of measurements and scales. Bundled up with her round glasses and red cheeks, she was like a benevolent grandmother, sitting upright at her table, examining slides under a microscope, freezing biological samples, slicing otoliths, and sampling various invertebrates, science that would later prove invaluable to Guido.

With the help of the Western anglers, Ksenya was able to create an updated picture of her steelhead. She was not happy with what she saw. In two short decades there had been a dramatic change in the steelhead population. The groups she had been monitoring in the seventies were substantially reduced, and many formerly identified variations were not present at all. The remarkable diversity that she had witnessed a few decades ago was greatly diminished. The question was why. The only human interference on these rivers was poaching. Ksenya had not imagined poaching to have become an epidemic of such proportions, but she was wrong. In the Russian Far East, poaching had become the most profitable and the blackest economy of all.

In the weeks to come, Guido watched it all quietly, spending time with his new Russian partners and helping to guide the American anglers. The fishing was stellar, though as a steelhead river, the Utkholok broke every rule he knew. In America, one would never find steelhead in peaty water. Steelhead liked cold, clear, strong rivers. To Guido, the Utkholok's slow, opaque water did not look promising—it looked like a meandering ditch that couldn't possibly yield a decent fish. But sure enough, the steelhead were there, swimming submerged like torpedoes, with only their fins breaking the water's surface.

It was Ksenya who explained that Russian steelhead didn't like warm water either, but they were retreating from an even icier body of water, the Sea of Okhotsk, which froze in winter. Steelhead wintered in the deep pools of these peaty rivers, coming in from the sea in the fall, when the temperatures started to drop. Guido saw that it made for an unusual fishing challenge.

More than the fishing and the river, Guido watched Pete Soverel, for it was with the ex–navy captain that his future lay. Pete ran his camp like a military operation; it was top-down, with Pete at the top. Guido learned much from the battle-hardened leader, but he was often treated like a subordinate officer. This rubbed him the wrong way, and made it tough to build the partnership between them. Guido was nonetheless struck by Pete's ability to lead and the iron-grip with which he exerted control. It was impressive and even instructive, but it didn't seem like the way to deal with Russians who, as far as Guido could tell, did what they could to make life difficult for Pete. Year after year, Pete and his cronies fought the Russian's assumption that they were in Kamchatka to spy, and wearily endured detainments and interrogations.

Pete's signed contracts were regularly reopened for renegotiation. For his regional contract, he was in the habit of paying up to 50 percent more than expected. There was no court, no legal recourse. Challenging the arbitrary laws of Kamchatka was not an option, but Pete went right up to the edge with his resistance, especially with the local outfitter they had contracted, who dropped into camp occasionally just to extort money. The outfitter had little to lose with such tactics; he didn't like Pete and would have been happy to run him out of Kamchatka. To Pete, a military man, such behavior was inexcusable; a deal was a deal. But arguing got Pete nowhere.

Guido appreciated that the world had been simpler thirty years ago, and for career soldiers like Pete, it had been clearer. There were allies and enemies; battles were either won or lost. Survival depended on control and command; problems were met head-on with force, and if your opponents did not yield, they were hit harder. It was a Cold War sensibility, and Pete's senior partners all shared it. That fall, Karpovich and Sager were both at the Utkholok camp, and the three of them generated the atmosphere of an off-duty military council. But they were commanders from a bygone era.

The real trouble was that Pete approached environmental protection like another naval battle. He took his position and dug in. The navy captain, who had a special loathing for bureaucrats, was inclined to sue rather than negotiate. For Guido, this intransigence presented the biggest obstacle of all. His own experience had taught him that flexibility and an open mind were the most effective strategies when it came to conservation. Of utmost importance was winning the cooperation of locals. This simple yet radical approach was one that no other international conservation group was embracing. Yet Guido knew that it was the locals who ultimately held the key to the future of their land, and if they did not see or feel the value of its protection, there was little an outsider could do to inspire their compliance. One had to talk to them, and more important, one had to listen.

By the end of his time on the Utkholok, Guido recognized that he and Pete had profound differences both as men and as leaders. While he did not yet know the local situation in Kamchatka, Guido could see that the Russians here were a proud and cynical people, and their long isolation had made them unreceptive to suggestions from outsiders. The locals did not trust foreigners, and they especially did not trust Pete. Guido assessed the situation. He first needed to distance himself from Pete, and then he needed the backing of a serious organization. The bold undertaking of protecting Russia's salmon strongholds required the support of a governing body that the Russians trusted. Only a few such organizations existed.

The United Nations had been watching Kamchatka with concern. The new Russia had birthed an economic class that was referred to as a kleptocracy for its genesis in the frenzied grab for control over what had once been vast state-run industries. The result was that a handful of men now controlled entire industries: gas, steel, railroads, ports, and aluminum. The management of Russia had fallen to the oligarchs, men who were not hindered by regulations or oversight, and could maximize their profits any way they saw fit.

The richest playing ground for expansion and profit was the Russian Far East, and the long-overlooked peninsula of Kamchatka, with its untold natural resources of oil, natural gas, mines of gold and silver, precious minerals, and timber. By the mid-nineties there was little doubt as to the direction Kamchatka would go if development proceeded unchecked. In Moscow, contracts were being drawn up for drilling and clear-cutting; there were plans for pipelines, and rivers were marked for hatcheries and dams. Suddenly it was open season on a land that had never seen modern man.

A few years earlier, in 1996, the United Nations and UNESCO had moved to bolster protection of the peninsula by officially recognizing the Volcanoes of Kamchatka as a World Heritage Site. It was a majestic expanse made up of six sites—soaring, snow-covered volcanoes that formed a group that represented "one of the most outstanding examples of the volcanic regions in the world." Once inducted, the site was protected by international treaty for its significance to the "collective of humanity." Now, as development escalated, the UN's environmental wing strove to work with Russia to strengthen and defend that site and others from rampant exploitation.

When word of the UN's Kamchatka initiative reached Guido, he immediately began investigating. Somewhere in this initiative was an opportunity, and he had to find it. He considered the site that had already secured international protection. He had seen these volcanoes from above, and the rivers that plunged from their heights. He had seen the forests and meadows and tundra, the bubbling hot springs and the cold, deep lakes. There were exceptional concentrations of eagles, bears, otters, foxes, and wolves, and untold species of plants, flowers, and trees. There were species here that lived nowhere else. Kamchatka's volcanoes were breathtakingly beautiful, but it was their biological health that made them valuable. Guido reasoned that such health would not have existed without a steady and regular infusion of the best food on earth—salmon. The entire food chain relied on salmon; salmon provided the sustenance, and were the lifeblood of the place.

Convinced of this fundamental reality, Guido went back to his Yale network and started asking around, looking for a way in to the UN initiative. It didn't take long for him to find a familiar name, and he could not have asked for a better one. Part of the small UN Kamchatka team was his second-year roommate, Jeff Griffin. They had lived together off campus on the second floor of an old Baptist church, where they were awakened on weekends with clapping and spirited "hallelujah"s. Jeff was a Montana boy, and like Guido he'd been raised fly-fishing. He was a dedicated conservationist respected in the field for his quiet charm, intelligence, and cultural sensitivity. It was little wonder he had excelled at the UN. Guido picked up the phone. He knew what his argument would be, and he hoped Jeff would get it quickly: without salmon, the nature preserve of Kamchatka wouldn't exist. The current designation of protection included mountains and lakes, but not the lowlands that held the biological riches the salmon had created. Guido would argue that the line had to be moved down from the mountains to include these floodplain rivers.

Guido knew that to get where he wanted to go, he needed the autonomy and the space to maneuver as only he could. He could not operate with Pete Soverel's team of Cold War warriors. He communicated as much to Pete, initiating a long and slow distancing process, and perhaps surprising the older man. But Pete was grasping that Guido had his own strategies, and would not take orders from him. In fact his new executive director would not answer to a chain of command at all; he would answer only to salmon.

CHAPTER 11

TIGER BY THE TAIL

W HEN GUIDO CAME BACK from the Utkholok, he faced the reality that he was an executive director without an organization. He had an old laptop, no office, and two months' salary. But one of his great talents was viewing things from afar and determining the next step. He decided that what he needed to do to get things under way with his new organization was to get a scientist of his own: he needed Xanthippe Augerot.

He had been thinking about Xan since he'd run into her at an American Fisheries Society conference in Sunriver, Oregon, some months earlier. She had presented a paper about the status of the salmon across the Pacific Rim, the result of years of research and the basis for her PhD. She had organized exchanges with Russian scientists to help to determine the health and distribution of salmon in the remote western Pacific Rim. In front of the crowded room, she brought up a crude geographic informational map that showed the whole arc, from Northern California all the way to Asia. Just as at Yale, she wore no-fuss clothes and had a no-fuss haircut. Her smile was wide and disarming as she greeted her audience and delivered a devastating message.

Xan reported that the data collected so far proved one inarguable discovery: salmon were disappearing from the southern ends of the rim's arc. On one end was a region Xan called WOCI—Washington, Oregon, California, and Idaho—and on the other end was Japan. She used a laser pointer to indicate the landmass in between, the Russian Far East, where 40 percent of the wild salmon population were thriving. One place in particular bore mentioning. The only place on earth known to host every species of Pacific salmon was Kamchatka.

Listening to her presentation, Guido felt excitement build inside him. The report was interesting, but he could see that it could be much more. Xan had colored the areas across the Pacific Rim that represented where the six species of wild salmon were abundant, and where their numbers had dwindled. Staring at the screen, Guido saw how Xan had translated a vast complexity into a simple, elegant picture. But what if she layered in more information—the location of hatcheries, for instance. They might be able to see if and how hatcheries affected wild populations of fish—or other factors, like dams and poachers. If properly built, Xan's map might show the massive salmon ecosystem of the Pacific Rim in its complex entirety.

On a practical level, Xan's map was a platform from which any number of stories could be told, and arguments formulated. If the complicated story of Pacific salmon could be made simple, people might understand, and if they understood, they could be inspired to support further understanding—and action. Guido felt in his bones that Xan's map was powerful enough to build an organization around—maybe even an entire conservation movement. But he needed Xan to make it happen. It was Xan who had the science he had never had; she had the language, the network, and the brainpower.

Xan and her team were broke; they'd gone as far as they could go with their academic funding. Guido didn't have any money either, but he was confident he could find it. He promised Xan he would raise the funds she needed to build her map out, allowing her to complete her research. He vowed to give the project the attention it deserved.

He went on to describe his stronghold vision, pointing out how Xan's research would complement it perfectly. Xan saw that Guido's plan was grand indeed, and that Guido was offering her a chance to tackle an enormous project for which she'd never had adequate support. Most important, Xan felt that Guido understood what she was trying to accomplish. When she said yes, Xan became the second staff member of Guido's Wild Salmon Center, which would quickly emerge as a lean, targeted machine with a scope not yet seen in the conservation world. Still, they needed to eat.

True to his style, Guido then went straight for the biggest fish he could find. He targeted Ted Turner, an idiosyncratic visionary who was notoriously deep in with the Russians. Turner was also America's biggest private-land owner, with many holdings dedicated to wilderness preservation. When Guido learned that Turner had just hired a man named Mike Finley to run his foundation, his hopes rose, for there was a good chance that Finley would appreciate the concept of strongholds. As chief superintendent of the country's major national parks, Finley had been responsible for the final phase of reintroducing wolves to Yellowstone, carrying himself the first acclimated Canadian wolf by mule-drawn sleigh into the Wyoming wilderness.

Turner owned four large ranches in the greater Yellowstone ecosystem. When word of Yellowstone's wolf project reached him, he asked Finley to personally show him the introduced wolves. Finley took Turner and his then wife, Jane Fonda, to observe the wolves in their acclimation pens, where the two groups of Canadian wolves were adapting to their new environs. The story of what had happened to the gray wolves of Yellowstone was a powerful lesson in ecology, Finley explained. When they were hunted into extinction nearly a century earlier, their disappearance led to a trophic cascade. Without wolves to curb their numbers, the elk of Yellowstone proliferated, and devoured the trees and shrubs along the park's creeks and rivers. Willow shrubs, used by beavers to build their dams, disappeared. Without a place to shelter and raise their young, beavers died off, and their dams went with them. Beaver dams had played an important role in the river ecosystem, slowing water flow with their thick clusters of rambling branches. Unchecked, the water gained momentum and began to cut into the land. Creeks ran faster and dug deeper into the earth. The new water level was too deep to nourish the shallow roots of willows and aspen trees. When these trees died out, songbirds and other plants dependent on the willow ecosystem went with them. The story of the wolves of Yellowstone had cracked open the story of ecosystems for the general public, and many people understood for the first time how interconnected the natural world could be.

After assisting in the feeding of the wolves, Turner asked Finley if he could show them Yellowstone's grizzly bears, which he had never seen. The three took off in Finley's jeep and spent the afternoon viewing bears and all manner of wilderness. Turner was so impressed with Finley's knowledge and initiative that he offered him a job at the end of the day. Finley declined, but Turner kept after him. "You have a deep love of nature and so do I," Turner said, adding, "We're going to be great friends." Sure enough, the two men explored the natural world together, traveling from rivers in Wyoming to Turner's ranches in South America. After turning down three more job offers by Turner, Finley finally accepted the fifth.

With Finley at the helm of Turner's foundation, Guido knew he had a chance at getting substantial support, for salmon and wolves had something essential in common. They were both keystone species. It was likely that Finley also understood that while bringing the wolves back to Yellowstone had allowed the ecosystem to begin repairing itself, the park would never be the same. The wolves of Yellowstone told a similar story to that of the wild salmon of the Pacific Northwest: a broken ecosystem could be mended, but it could not be fully restored.

Resting on the gentle shoulder below the peak of Mount Hood was the stately Timberline Lodge, a grand, rambling structure built in the thirties by the Works Progress Administration. A ski resort in winter, Timberline hosted meetings and conferences in the off-season, one prosaically titled the Consultative Group on Biological Diversity.

Mike Finley was in the audience when Guido took the lectern to describe the importance of salmon strongholds, and how even billions of dollars could not correct the imbalances of compromised river ecosystems. He had by now become a skilled speaker and knew just how to connect with an audience. Finley approached Guido afterward and told him he'd enjoyed the talk, and agreed wholeheartedly with the stronghold argument. The two men took an instant liking to each other. Guido instinctively trusted Finley, and over lunch he took the opportunity to ply Finley with questions about his boss. Finley knew what Guido was after and was happy to oblige. First off, he told Guido, Turner didn't like long meetings. Finley had learned this early, when a particularly lengthy meeting was interrupted by a loud thumping sound, which turned out to be Turner's head banging against the table in bored frustration. Second, Turner was a man who got things done. He played by his own rules, or he made up new ones. He had two pressing causes: protecting the environment and getting rid of nukes. Like Guido, Turner was focused on Russia.

During the Cold War, Turner had conducted a one-man diplomatic mission to ease tensions between the two superpowers. The longstanding threat of nuclear war served no one and had to end. It was absurd that politics had trumped the Olympic Games for an entire decade. Turner balked at the unnecessary waste and loss of opportunity for camaraderie and answered by launching the Goodwill Games, an international sporting event on an Olympic scale that would be kicked off in Moscow. Turner lobbied hard for America to participate, and after eleven years of noncompetition, American and Russian athletes played together once more, though the Russians came away with more medals. The games did much for Russian spirit and national pride, and the country was indebted to Turner and his seemingly bottomless pockets. When he visited Moscow, Turner was treated with the respect of a dignitary, and was escorted by top officials—including a high-ranking KGB officer named Vladimir Putin.

As far as Finley knew, his boss had not invested significantly in Russia's environment. The key, he told Guido, would be getting

Turner to Kamchatka. Turner was drawn to action and adventure, and he was an avid fisherman—he'd probably jump at the chance to fly around in old Soviet Mi-8s with one of his four girlfriends. In the meantime, Finley affirmed the Turner Foundation's support for the Wild Salmon Center and pledged $100,000, a gift that would continue for years to come.

When Turner met Guido later that year, the media tycoon judged Guido as quickly and accurately as he'd judged Finley. Guido was another straight-talking naturalist who knew his subject, and was serious about taking action. Strongholds made sense to Turner, and Kamchatka was damned intriguing. When Guido invited Turner to come fish its rivers, Turner readily accepted. Years later, in 2008, Turner joined Guido, escorting one of his girlfriends to Kamchatka, where he held forth around the campfire, and cast his fly into Russian waters.

With Turner's support, Guido was in a position to hire his third employee—the one person who could help him make the United Nations project feasible. It was early morning in Magadan, but Misha Skopets picked up the phone. Guido launched right in. Would Misha accept a position with the Wild Salmon Center? His job would be to explore the rivers of the Russian Far East and make assessments of their health in terms of both salmon abundance and diversity. They needed to make a list, Guido said, of the best salmon rivers in Kamchatka. Then they needed to prioritize their protection. Misha understood, and accepted the assignment instantly. That week he packed his bags for parts of the Russian Far East even he hadn't seen.

Guido was proceeding with the assumption that the United Nations gig would be his, but there was still legwork to do. Jeff Griffin had agreed that the key would be in establishing the link between salmon parks and the health of the greater ecosystem, which, for Guido, wouldn't be a problem. These fish were the reason for the beautiful and robust habitat of the UN's World Heritage Site. Kamchatka's riv-

ers represented the greatest salmon stronghold in the Pacific Rim—
shouldn't they consider protecting them?

Guido made the bold recommendation that a separate salmon program be created for him to run and argued his case persuasively
enough to win an official consultancy to the United Nations. But
there was a hitch. While the UN could get him into Russia, Guido
would have to cover the costs of whatever he planned to do there. He
was playing at the big table now, and he needed a tall stack of chips.
Guido wasn't close to having the money; Turner's donation barely
covered three modest salaries. He had been hammering away at various donors for moderate amounts of $10,000 or $15,000 to support
his stronghold plan, but even if all these donors came through, he still
wouldn't have enough to get a major program going in Kamchatka.

It was then that Peter Seligmann gave him another deceptively simple piece of advice that he would return to again and again. Be crystal
clear about your mission, decide exactly what you need to get it done,
and find the right person to back you. What would it take to get the
stronghold job done in Russia? Peter asked. Guido should take out
his legal pad and consider every aspect of the project, penciling in
numbers for everything from the nuts and bolts of the Wild Salmon
Center operations to the grand unknowables of their objectives in the
Russian Far East. Once he came up with a number, Guido would
have to summon the courage to ask for it, and he would have to ask
the right person. The strategy spoke to Guido's preference for removing unnecessary steps and collapsing the distance between where he
was and where he wanted to be.

Before he hung up the phone, Peter passed along a name. "Here's
a place to start," he told Guido. "You're going to owe me for this."
Guido wrote down the name, which meant nothing to him, though he
could hear Peter smiling over the phone. "Get ready for your world
to change," Peter said.

Gordon Moore had started a company called Intel. Guido knew of
the Silicon Valley giant, though the high-tech world seemed distant to
environmental causes. Peter assured him that Gordon Moore was un-

usual; he had a prodigious intellect and was aware of the dangers facing the earth's habitat. He'd been concerned enough to join Peter's board at Conservation International, where he'd proven to be a valuable member. Gordon and his wife, Betty, had a family foundation, but to Peter's knowledge, they had never funded projects in Russia. Guido was welcome to knock on the door and try his luck. Peter thought he had a chance. Gordon was a science geek, and deeply curious about the nature of the world. And, Peter added, he was a fly fisherman.

Guido immediately drafted a letter of introduction to the Moore Family Foundation. A few days later, he followed up with a grant proposal describing his stronghold approach and the urgent need to fund the upcoming UN project. The following autumn, the United Nations Development Program, or UNDP, would be holding its initial meetings in Petropavlovsk. There was much work to do before then.

In the meantime, Guido was preparing for an expedition he found deeply intriguing. In their explorations of the rivers of western Kamchatka, he and Pete Soverel had come upon the Krutogorova River. North of the Utkholok, the Krutogorova yielded an enormous number of fish, but, otherwise, very little was known about it. The river was difficult to access; what lay beyond the mouth was a total mystery. Guido and Pete learned they had good reason to try to get there, and as soon as possible. The Krutogorova had been designated as the terminus for a natural-gas pipeline that would bore into the ground at the river's mouth and run down the Kamchatka peninsula, all the way to the Bolshaya River. Excavation had already begun, which meant the mouth of the river would soon be gouged, mud-filled, and obstructed by equipment. The Krutogorova's pristine habitat was about to radically change, and with it, the population of fish that spawned there.

Guido and Pete had decided to float the river from its headwaters to the mouth and see what they could discover about the causes for the river's exceptional productivity. The hope was that the Krutogorova would tell them something essential about the nature of salmon and the salmon ecosystem. There was something else that made the trip potentially exciting. He and Pete had recently stumbled upon a man who provided a piece of the puzzle Guido hadn't known was missing. They hadn't been looking for Jack Stanford, but Stanford had found them. A maverick American river biologist who had spent his life studying rivers, Stanford had heard about the Wild Salmon Center and its work in Kamchatka. He'd gotten wind of the expedition to the Krutogorova and approached Guido and Pete about coming along. Stanford had many reasons for wanting to see such a river, and he and his colleague offered to pay their own way. Guido was happy to have the American scientists along; who knew what connections they would make. And so Professor Jack Stanford and PhD student Nick Gayeski were given two seats on the float trip down the Krutogorova. Their discoveries would make it the most scientifically important expedition to date for the Wild Salmon Center. It would also be one of the most dangerous.

CHAPTER 12

THE KRUTOGOROVA

O N THE TARMAC AT Petropavlovsk's Yelizovo Airport in Kam-chatka, Russia, a group of eighteen Russians and Americans wandered, jet-lagged, among stacks of scientific equipment, fishing rods, food, and miscellaneous gear. It was mid-September, 1999, and the Siberian summer would be over in a matter of weeks. Fallen leaves swirled in the occasional cold gust, a harbinger of the long, stormy winter to come. Moscow had sent their senior scientists Pavlov and Savvaitova, and a few junior scientists. Sergei Pavlov, the son of Dmi-tri, was one of them. Pete and Guido had invited some sponsoring fishermen from America and, for the first time, two American scien-tists. One, Jack Stanford, had some years earlier presented findings to the scientific community that changed how they understood rivers forever.

Jack Stanford had grown up on a ranch in Montana, fishing on trout streams. As a river ecologist and ecology professor at the Uni-versity of Montana, he had spent most of his life in the wilderness, where he had made some startling discoveries. It was 1973, the same year that Savvaitova had published her "Noble Trouts," that Stanford had made his most radical find. He had been studying stone flies at his bio-station in northern Montana. These big flies were a critical organism in the river habitat. Born on the river bottom, they crawled out to become airborne as adults. In observing the flies, Stanford had come across something that stumped him. Of the dozen species of stone flies he was documenting, there were eight he could not find in larval form. He saw them flying around in the air but couldn't locate them in the river, in their larval form. It was as if they appeared out of nowhere.

One day, a local water engineer came to Stanford's lab at the Flathead Lake bio-station with a bottle full of murky water he'd collected from one of his buried pipes. In the bottle were the larvae Stanford had been searching for. Peering into the cloudy bottle, Stanford had an epiphany that changed his career. These flies, found in pipes far from the river, had adapted to life underground. All of his life, Stanford had been getting in and out of rivers, assuming they started and ended with their banks. But these stone flies told a different story.

This discovery, made miles from the river, hinted at a hitherto unknown underground ecosystem. Stanford went on to pump up more stone-fly larvae from deep underground, where they had been living in the aquifer. Instead of maturing on the river bottom, these enigmatic species traveled down underneath it. Stanford spent the next decade establishing that in the network of streams extending far below and beyond the main channel of the river was a third dimension. Water in a healthy watershed didn't flow just longitudinally from mountain ridges or laterally with flooding—it also flowed vertically. Invisible from the surface was a dynamic interplay between river water and alluvial aquifer that both cleaned the water and created myriad microhabitats.

After publishing major articles in both *Nature* and *Science* magazines, Stanford applied his findings to the most valuable rivers in the Pacific Rim—salmon rivers. Here, the third dimension changed everything. The prevailing wisdom of Western fish scientists was that salmon species populated different parts of rivers and rarely commingled in any of them. It made some sense; each species had specific needs—particular water flows and temperatures and gravel sizes for their egg beds—and these needs were met in different parts of the river and, in some cases, in different rivers altogether. Tasked with the protection and rehabilitation of salmon, conservationists and fish managers designed programs and systems based on the assumption that salmon species were more or less river specific.

But for Stanford, this prevailing wisdom no longer made sense. In fact, it was wrongheaded. Under the river and stretching out into the aquifer, he had found a sufficiently diverse and complex habitat to host all species—and it was built in naturally. The trouble was, fifty to a hundred years of logging, dams, and agricultural interference had nearly destroyed this habitat.

Stanford was no longer in doubt that most North American rivers hosted only two or three species of salmon because they were too degraded to host more. The trouble was, there was no way to prove the inverse on such compromised rivers. It would be like trying to characterize the health of someone with a wrecked immune system. On these rivers, millions of dollars of failed restoration attempts had shown that there was no way to turn back time. Perhaps, Stanford thought, the water above the ground had once mirrored the water below. Before it had been restricted to a single artery, the river might have flowed through many veins and capillaries, circulating through fields and even forests. Stanford thought he might have glimpsed a clue about how America's rivers had once been. It was like studying a lost world.

He continued to be haunted by this phantom ecosystem, feeling it held answers to questions that could direct the protection and restoration of the salmon rivers of the Pacific Rim. As his research progressed, Stanford couldn't help himself from developing theories that were impossible to prove. When he learned about the trip to the Krutogorova, he didn't hesitate to jump on board.

Pete and Guido split the expedition into two groups. Pete's group would establish a base camp near the mouth of the river where the

American anglers could fish for steelhead and the Russian scientists could set up their field lab. Guido's group, meanwhile—the float team—would fly as close as safely possible to the river's headwaters, sixty miles to the east, and raft back through unexplored tundra.

Under Stanford's guidance, the team would work together to collect samples of water, gravel, silt, vegetation, and fish while they walked transects across the floodplain, treading in a succession of straight lines across the riverbed and measuring the organisms they found at regular intervals. With instruments that recorded and described the terrain they covered, they would attempt to map the features of the river landscape, the likes of which the American scientists had never seen. This map would serve as their first baseline for an untouched watershed.

As always with Russian bush helicopters, preparations were rushed. Now, as the crew hurried to sort and pack the gear, the overbooked pilots were anxious to be off. The plan was to fly both helicopters three hundred miles northwest to base camp. Then, after a final but all-important gear check, the float team would be choppered to the headwaters.

An unspoken tension ran through the eight members that made up the float team. While most of them were highly educated and experienced in wilderness survival, they had no idea what to expect on this river. There were no photographs or travel accounts, only old Soviet maps that showed scant detail of the interior.

Stanford was mildly concerned. He had spent much of his life navigating rivers, but the rivers he knew had been tamed by agriculture and hemmed in by development. They had stable banks; one knew where they started and stopped, how much water flowed through them, and how it flowed. But he had no data about the Krutogorova.

He asked one of the Russian guides to translate a message to the pilot. "Please fly low over the river, so we can scout for rapids, falls, and other impasses." The pilot had nodded in understanding at the American's request—*"Da, da, da . . ."* Stanford couldn't have known that from a Russian helicopter pilot, "Yes, yes, yes" was an unfavorable response. *"Da"* meant the thing would get done. *"Da, da"* was less likely. *"Da, da, da"* meant forget about it—they wouldn't even consider doing it.

As the choppers filled with equipment, Guido looked around for the requisite laika—compact, muscular huskies bred in the remote Lake Baikal region for their particular aggression toward bears. These dogs were essential for venturing into the interior of Kamchatka, home to the world's largest brown bear population. Laikas chased the enormous predators across rivers and into forests, barking and harassing them along the way. It gave a laika great satisfaction to chase a bear; the farther the dog could make the bear run, the better.

Guido had been protected by laikas before. But he hadn't seen one among the Russian crew, which made him uneasy. It was when the helicopter was almost packed and the engines started churning that a lone miniature pinscher jumped on board, after the cook. Guido stared incredulously at the tiny, fine-boned canine and murmured to his coleader, Sergei Pavlov, "Don't tell me that's our bear dog." Pavlov nodded as the doors closed.

Stanford was not as perturbed by an unexplored river as he was by the Russian Mi-8 with its peeling paint and black streaks of oil running down its hull. Inside he was alarmed to find no seats, and spare fuel tanks just sitting there, not even strapped down. It was with real apprehension that he and his colleague, Nick Gayeski, hoisted themselves into the helicopter. Sitting against the cold metal of the interior, Stanford eyed the diminutive bear dog across the cargo hold, looked at Gayeski, and said, "We're going to die."

The old Soviet helicopters had flown many miles. Their parts had been switched out and replaced so often that there was little left remaining of the originals. But, Guido assured him, they were incredibly sturdy machines. The engine turbines came to life with a whine, followed by the deep throbbing of engaging rotors that increased in speed until the whole helicopter shuddered. The smell of smoke and kerosene filled the air, and as the noise grew to a deafening roar, they lifted off.

It was too noisy to talk, but Guido and Stanford were fully occupied with watching the scenery and snapping photographs of the terrain below. Three hours into the trip, they were roused from their thoughts as the two helicopters inexplicably split up. They watched as Pete's chopper went one way and theirs veered east, away from base camp.

At first, Stanford thought the helicopter pilot had not understood his request to fly low over the river to be dropped off at base camp. As it was, they seemed to be taking a beeline to the headwaters. Guido watched as Stanford gestured to Pavlov and pointed emphatically to the map. This had not been the plan. The two men huddled and tried to locate where they were in relation to the Krutogorova. The pilot offered no help at all, and was unreachable between his headphones and the roar of the engines. In any case, Russian bush pilots answered to no one. Out the window, Guido noticed they had started to fly through clouds. Below he could see bits and pieces of a river with many channels.

When the Krutogorova finally came into view, a quiet panic ensued. The pilot had flown too far upriver. They were in the fast water of the foothills, where the river came crashing down from the mountains. What they saw was not one river but five or six rivers crisscrossing through a dense cottonwood forest. Fallen trees had created blocks around which the water had parted or jumped, creating a complex series of braided channels.

Stanford yelled to Pavlov, who pushed into the cockpit and ordered the pilot to land at the first possible location. At the first decent riverside clearing, the chopper touched down in the middle of a glacial tundra. The men stepped out of the helicopter and looked around. It was as if they had entered a Pleistocene world. The ground underfoot was soft with mosses and the air fecund with growth. Nourished by the nutrients of decaying salmon, the vegetation was overgrown and immense, and cottonwoods had trunks the size of Douglas firs.

The chum and pink salmon were at the tail end of their summer run. Recent high waters had left dead fish hanging in the low branches of trees. Brightly colored carcasses winked like shards of glass scattered across the floodplain.

This preview of the river was met with silence. The good news was that there were clearly masses of fish. The bad news was that the transporters of the salmon across the floodplain were bears. Equally bad were the giant clumps of fallen old-growth trees and the confluence of rivers clogged with logs.

Stanford had been floating rivers all his life; he had yet to encounter a river he couldn't navigate. The fact was, the Krutogorova was structured in a way he had never seen—and it was full of hulking obstacles. As anyone who floated rivers understood, little was more perilous for a floating object than the force of a river diving under an immovable blockage. No human being or raft could withstand the pressure and would either buckle, break, or be sucked under. Rescue was next to impossible. On Russian rivers a chain saw was a requirement, and they had made sure to pack one on this trip.

The Russian pilot did not linger. As soon as the gear was pushed off the chopper, he fired up his engines. As the rotors started to turn, the men threw themselves on top of the gear to keep it from blowing away. The tremendous whir of the Mi-8 faded into silence as the group stared after it, slightly stunned. The four rafts, oars, life jackets, tents, food, and equipment had been dumped on the rocky beach without the requisite base-camp check on equipment, radios, and other necessary items that would help them survive the days to come.

Most critical was the radio, which was located first by Sergei #2, their Russian outfitter, cook, and owner of the pinscher. Prearranged twice-daily check-ins with base camp were protocol. It was important to let Pete know they had arrived, and that the river looked dangerous. When no sound issued from the radio, Sergei fiddled unsuccessfully with the dials and eventually stowed it away.

Guido joined Stanford at the water's edge as Stanford somberly studied the Krutogorova. He told Guido it was better—and worse— than he had imagined. Each braid presented an option; if they chose the wrong braid they could end up in a logjam that could be up to a hundred feet long. If they got caught in the water underneath one of these there would be zero chance of survival. Nor did they want to get sucked into the whirlpool at the front end of a jam. What he didn't need to say was that out here, there was no way to get help. An injury, even a simple gash or sprain, could prove disastrous.

They would not have to deal with the river quite yet. The plan was to make camp and stay for a few days while they recovered from their jet lag and explored their extraordinary new environs. Stanford was itching to march into the unknown, and Guido was anxious to go with him. Guido had already learned things from Stanford, who pointed now at something he found intriguing. Cottonwoods and willows, trees that did not like cold weather, were flourishing in the middle of the tundra. Stanford suspected that their roots might be warmed by a giant alluvial aquifer.

That afternoon, Stanford, Gayeski, and Guido walked the first transect. They could only take so many samples out of the country, so they supplemented the hard science with photographs and drawings. Stanford took the time to sketch the remarkable vegetation. Below the canopy of willows and cottonwoods was an understory of cow parsley that was thicker and higher than a cornfield in Iowa and growing so fast he could practically hear it. Stanford said that these plants were loaded with protein from decaying salmon in the soil and provided a supercharged diet for insects, bears, and whatever else lived on the river.

Stanford was beyond excited. To his knowledge, no ecologist had fully described the role of salmon in the river ecosystem. They had, however, found a major indicator. The easily traceable nitrogen isotope NI5, unique to marine systems—the ocean—had been discovered in trees along the banks of salmon rivers and deep into the interior. NI5 had even been found in 2,000-year-old trees. The evidence pointed to salmon being the primary generators of their ecosystem—but no one had yet ventured such a hypothesis. Looking around, Stanford thought he had never seen such incontrovertible proof.

As they made their way across the floodplain, the three men conducted a geomorphic study, making gradient measurements of the
land's slope with a stadia rod, a navigation tool used for centuries on
sailing ships. It was essentially a small telescope with a sighting level
resting on top of a tripod that they positioned periodically as they
walked, Gayeski holding the scope as vertical as possible while Stanford read the measurements. When they had gathered these measurements all the way to base camp, they would have an idea of how
quickly the water dropped from the foothills to the sea, a significant
element of the river's "map."

After the precipitous fall from the mountains, the gradient of the
Krutogorova was relatively flat. It was here that rivers and their maze
of channels spread across the wide valley, creating ribbons of foliage
and forest in the middle of the tundra. It was a landscape Stanford
had only imagined—but he had imagined it correctly. He and Gayeski
waded through the clear water of side channels, shouting out like
boys as hundreds of juvenile salmon swam between their legs. Guido
stayed close to Stanford's elbow, not wanting to miss anything. As an
angler, Guido thought he had a good grasp of rivers, but Stanford
was pushing his understanding to a whole new level.

They returned to camp to begin the second phase of their research,
one that Guido could easily contribute to. In order to see what was
swimming in the river, they had to catch it. Sergei Pavlov had already
headed off with a spinning rod. He and American biologist Don
Proebstel were conducting research on arctic char and were not as
interested in the sport of fishing as they were in procuring "lethal"
samples of char, which they would dissect in the evening by the fire,
storing genetic samples in a canister of liquid nitrogen.

Stanford had tasked Guido with catching the elusive Kamchatka
rainbow trout, a powerful and spirited fish that ichthyologists knew
little about. Like everything in Kamchatka, rainbow trout were bigger and stronger than their American counterparts. Along with flies,
they also fed on small rodents.

Guido glanced at the sky. It was almost the witching hour, when
fish started rising to the surface to feed. He headed in the opposite

direction from Pavlov to find his own place in the river. Positioning himself above the drooping branches of an alder tree, he took in the currents, the trees, and the insects in the air. The sun disappeared behind a thick cumulus cloud and the light began to fade.

The river was bursting with life, and with numerous signs of bear, Guido's senses were dialed up; he was acutely aware of his place in the food chain. Still, it was thrilling to be in a river that had never seen a dry fly. He studied the shadowless water and considered what to tie on the end of his line. On the square patch of sheep fluff on his vest nestled a selection of hand-tied flies, but these rainbows had been known to feed on mice. Guido removed a box of larger flies from the pocket of his vest and selected a mouse fly he had tied for this purpose out of deer hair and a strip of rabbit hide. The unwieldy fly required a curious and challenging technique. Instead of floating like an insect on the water, this fly had to imitate a mouse swimming against the current as it ferried across the river.

Guido cast above the water and made a quick mend so the fly was not swept too quickly downstream. He then raised his rod tip slightly and wiggled it so the fly mimicked a swimming mouse dog-paddling with its head out of the water. He had fixed his eye on a calm stretch of water on the opposite bank where he could almost feel the presence of big fish. He "swam" his mouse across the smooth water, methodically working his casts downriver, never doubling back, doing nothing to spook whatever lay beneath the surface.

It wasn't long before he saw a rainbow approach his mouse. The large, dark shape followed the fly underwater, its back cresting out of the water before it struck. When the water exploded, Guido quickly lifted his rod. The strike of a big rainbow is violent and fast, and he had but a split second to secure his connection to the fish. Setting the hook, he experienced the familiar rush of being joined by the thinnest of lines to a fierceness and vitality that ran through him like a current. His reel clicked as the fish took off downriver, leaping once and then twice, its powerful mass rising and flashing in midair before disappearing.

The trout fought well. It took Guido some time to pull it in, never rushing and never allowing enough slack for the fish to wrap around a rock or reach the tangle of trees on the opposite bank. The rainbow's scales were iridescent, and it glittered below the surface. As the fish rolled toward him, Guido backed into calmer water and reached over his shoulder for his net. He didn't like to take long with the science. Every moment was a moment the fish needed to be somewhere else. While the trout rested, spent, in his net, Guido made quick work of measuring, scaling, and tagging before tipping the fish out of the net and watching its gleaming girth slip quietly back into the river. He recorded the buck's size at six pounds and twenty-five inches. A big rainbow, by any standard.

Back at camp, Guido related the details of his catch to Stanford, whose ebullient mood was contagious. After recounting the dramatic moments of the day, they changed out of their waders into warm clothes and joined the others.

The Russians had outdone themselves preparing the introductory dinner. The folding table held plates piled with black bread and salmon caviar. There were platters of sausages and a pot of borscht. As dinner began, vodka was poured into small glasses that would be upended with every toast. The first one, delivered by Sergei Pavlov, followed a protocol Guido was already familiar with. It started with a humorous story about a beautiful woman—Tanya in the *banya,* in this case—and proceeded more soberly to honor scientists who had gone before, as well as the greatness of the Russian Academy of Sciences and the invaluable role of research. A glass was finally raised to the noble fish that swam in these mighty Russian rivers, rivers mighty enough to bring former enemies together, making them all one before the mysteries of science and nature.

Stanford laughed and tipped back his vodka. The Westerner's friendly manner and cowboy drawl were as easy as his constitution was tough. Guido noticed that the Russians were warming to him. He was, perhaps, a type of man they recognized: strong and rugged and completely at home in the wilderness.

When it was their turn to toast, Stanford and Guido offered their amateur best, striking a more plebeian note as they raised their glasses to their good luck at being in such a perfect place, especially after they had screwed up their own salmon habitat. The two Russians nodded sagely, warmed by this admission of weakness. By the time the group fell into their tents, their bond was strengthened by a mission that transcended both history and politics.

The second day, Stanford, Gayeski, and Guido walked another transect and jubilantly counted three species of juvenile salmon: coho, chinook, and cherry. If they found three species coexisting this high in the river, how many would they find on the wide, complex transects at base camp?

CHAPTER 13

AT THE RIVER'S BOTTOM

ON DAY THREE, it was time to go. Sergei Pavlov had mapped the trip out, choosing campsites at intervals that would get them to base camp on time. He instructed half the group to start blowing up the rafts with a hand pump and the other to pack the gear.

Stanford noticed right away that the ropes on the boats were askew. The painter ropes were oriented on the downstream side of the boat and should have been on the upstream side. He flagged the worrisome detail to Guido and wondered if the Russians knew what they were doing. Meanwhile, Sergei had found the chain saw, but not the gas required to run it. Soon he requested an all-out search for the red plastic container. After a flurried hunt, they realized they had lots of vodka, but no gas. What they had against the giant logs was a hatchet and a single plastic-handled handsaw. They also had whistles to alarm in the case of bears.

At this point, Stanford effortlessly assumed leadership. He gathered the group and spoke to them plainly, saying he thought he could get them down the river, but that they would have to listen to him. Once on the river, they would need to pay attention to his whistle, which he would use to both direct and alert them over the noise of rapids. The whistles were distributed and the men were assigned to boats. The Russians would man the two large supply rafts, while the Americans took the more maneuverable smaller rafts that were better suited to fishing and gathering samples. Stanford suggested that he and Gayeski go first in their small raft so he could "read" the river for the following boats. The moment they pushed off from shore, the rafts were swept from the bank.

Guido understood the braided floodplain river better than when he'd floated the Bolshaya with Misha. Stanford had explained that an untamed river ran down the floodplain like the tail of a whip, curving to one side until it met resistance that curved it to the other. As the water coursed to the sea unobstructed, sliding from one side of the valley to the other, it chewed into the soft dirt of the riverbank, washing away the root covering of trees until the trees fell into the water, creating logjams. At the jams, the water rose, backed up, and jumped the main channel, diverting water into a side channel. For every river mile there were at least ten smaller channels weaving laterally through the floodplain, which would be the primary challenge to floating it. To avoid deadly logjams, they had to choose the right channel.

Immediately after rounding the first bend the rafters dove into a forest, where they ducked to avoid overhanging branches that restricted their vision to a scant few yards. Then the river started dividing into smaller and smaller braids. As they were whisked along, Stanford counted the channel separations carefully, noting when the channel they had chosen received water from another channel or split off on its own. He blew his whistle as he made choices, waving the other rafts to steer to the left or right braid. In his notebook he marked which channels had established plant life, indicating older, more established braids. The rafts floated one by one down channels clogged with vegetation. When they couldn't see him, they could hear Stanford yell "right" or "left" as the rafters behind grabbed branches to slow themselves down so they wouldn't miss the correct channel.

There was little talk as the men registered weeds and grasses pounded flat by bears and paw prints in the muddy shallows. They

rowed vigorously to avoid logs and other obstructions that could easily overturn their rafts, blindly following Stanford's yelled instructions as they shot down the gauntlet of obstacles. A few miles along,
the river widened and slowed, and, resting their oars, the men sat
back as the water grew calm and the scenery around them opened up.
Stanford pointed out natural chunks of coal along the gravel beds, an
indication of mineral deposits below. Above them, a Steller's sea eagle
sat perched in a cottonwood like a gargoyle, watching them with its
sharp yellow eyes. The largest eagle in the world, it had a fierce
hooked beak and talons strong enough to carry a large salmon.

The brief respite ended when the men in the lead boat waved a
warning as they disappeared around the next bend. Rounding the
corner, the following rafts came face-to-face with three bears feeding
on the riverbank, a sow with two cubs. Stanford smiled and shook his
head. He had never seen bears so big. They looked three times as big
as Montana grizzlies—but Montana bears didn't have salmon on the
menu. The sow was easily a thousand pounds of fish-fed hulk and
stood a full nine feet tall as she took a long look at the rafters. The
men floated silently toward her, helpless to create a respectful distance. They were dangerously close when the sow turned and disappeared into the grass, the two cubs in tow.

Back at base camp, Pete had cause for worry. There had been no
word from Guido. He knew exploratory float trips presented heightened safety concerns. The float-team leadership had to be extremely
competent, resourceful, and careful. In the event of a medical emergency, they'd be in trouble. The Russians didn't allow either GPS
devices or satellite phones to be brought into Kamchatka. Even a
working radio put them at least twenty-four hours away from any
outside assistance. For the float team, the simple act of stepping in
and out of a boat required vigilance, as careless inattention could result in a broken arm or ankle—an inconvenience in other circumstances but a serious issue in Kamchatka. At check-in time, Pete tried
the radio again with no success.

Increasing his stress was an unwanted guest. An inspector for the Russian Federation's Ministry of Natural Resources and Environment had arrived and announced that he would be camping out with them for the duration. The official, a buffoon who seemed to understand very little about the environment, watched them closely. He then proceeded to do whatever he wished—smoking, drinking, and at one point taking one of Pete's jet boats without permission for a joyride on the river. Pete watched from the shore as the inspector tore around, maintaining the barest control and, incredibly, failing to either kill himself or sink the boat. If this was the future of Russian environmentalism, they were all doomed.

Pete turned his attention to the fish. If they could find steelhead, all would be well. While the American anglers settled into their daily routine, Ksenya Savvaitova and Dmitri Pavlov assembled their makeshift laboratory and prepared to analyze their allowance of thirty lethal samples of fish. Savvaitova was focused on the question she still hadn't answered: What drove steelhead choices about whether to stay in the river or to leave, and when to leave—was it something in the ecosystem that was cueing the fish? Cheerful and energetic, she worked late into the night studying otolith bones that she extracted from the center of a fish brain by splitting the fish head with a large, sharp knife.

The two Russian ichthyologists anxiously awaited the arrival of the float team, curious to know what the American river biologist had observed upriver. They didn't quite buy Stanford's theory about the floodplain; they had never witnessed this so-called third dimension of the river. Still, they were open to collaborating on anything that would help support the future of their joint scientific enterprise.

Forty miles upriver and on day three of radio silence, the float team had lost precious time. They had been forced to drag their rafts around five logjams and were finally making progress. But now Stanford was leaning over the front of his raft and staring through the water at the substrata and cobble ties on the river's bottom. The pebbles and sediment were getting smaller and smaller, which meant that something ahead was stopping the water's flow, something they couldn't yet see. A moment later, his whistle pierced the air.

The boats behind watched as he rowed frantically for the bank, grabbing branches to keep his raft from going any farther. The other three rafts scrambled to follow him, slapping the water with their oars and fighting the current that threatened to take them past Stanford and whatever he was trying to avoid. Once onshore and clinging to branches, they tied up to one another. In front of them was a channel-spanning tree that was impassable by water or land—but this was not the obstacle Stanford had read in the river. There was something bigger coming. The two Sergeis wordlessly dug out the hatchet and handsaw and started chopping away at the trunk of the tree while the Americans looked on.

After an hour and a half of chipping and sawing, the cottonwood dropped, and when the rafts pushed off again, Stanford and Gayeski took the lead as the river picked up speed and made a sweeping bend to the left. Over a quarter mile later, it bent back to the right. It was then that Stanford saw the real obstruction. The river was split by an island. The smaller, farther channel flowed to the right, while the channel they were following flowed into a massive logjam.

An eighth of a mile above the jam were two large fallen cottonwoods that lay like sleeping giants, one on top of the other. Stanford steered to avoid them. Past the trees they were relieved to see a gravel spit about five hundred yards above the jam, and they rowed hard for it. Here they pulled their raft out of the river and waited for the other boats. It would take time to strategize about how to deal with the jam below.

Half a mile upriver and out of view, an unlucky chain of events was unfolding. The three remaining rafts had been parallel-parked, one behind the other along the bank. When Proebstel decided to launch his raft from the rear, the current caught it too quickly and it bumped into the next raft. Proebstel's bowline got tangled around an oar, spinning his boat and slamming it into the next boat in line, Sergei's big supply boat. The jolt was enough to yank the painter rope from Sergei's hand, and suddenly two boats were being swept downstream, one of them unmanned. Without being able to see Stanford or what was coming up ahead, Proebstel attempted to catch the swifter, heavier supply boat. Rowing strenuously, he managed to get ahold of it briefly, but just as soon had to let it go as the current bucked it out of reach.

Downriver on the gravel spit, Stanford was relieving himself in the bushes while Gayeski positioned himself to receive the other boats. Gayeski froze as the supply boat, piled high with the expedition's food and gear, appeared from around the corner—manned only by the pinscher. Following was the chase raft, woefully far behind. The only obstacle between the supply raft and the logjam downriver were the two fallen cottonwoods.

Thinking fast, Gayeski ran to the bank, made his way to the submerged trees, and waded into the water. Here, the raft would come straight to him. While he wouldn't be able to hang on to the boat for long, if he could find its hundred-foot bowline, he might be able to throw a couple of loops of rope over the tree and hold the raft until Proebstel caught up.

Gayeski was knee-deep in the river and positioned in relative stability when the raft came into reach. Then, a few yards above him, it lurched to a dead halt. Gayeski stared, confused, as the raft strained against the current. The rope had caught on something underwater—something strong enough to hold the boat in precarious suspension. Buffeted by oncoming waves, the raft would only hold for so long. It was then that Gayeski noticed a small brown head bobbing in the water. Somehow the dog was now in the river. Looking closer, Gayeski saw that the pinscher was attached to the raft by a short leash that was pulling him under the raft. Their bear protection was drowning.

Pitching forward, Gayeski threw his torso onto the raft, grabbing onto the side with one hand and groping below for the leash with the other. Grasping the leash, Gayeski dragged the half-drowned pinscher from the water and heaved the dog back on board. In the same moment, the raft lurched free of its obstacle and began to drift again, now heading straight for the cottonwoods. Gayeski slid back into the water and vainly held on to the downriver side of the raft, crow-hopping downstream as the river carried them both toward the trees. The water was getting deeper, and Gayeski was badly positioned between the raft and the trees. He was now waist-deep in water and he knew he was in trouble.

As they approached the cottonwoods, Gayeski's options were grim. He eyed the water plunging under the submerged trees and wondered if there was enough space between their trunks and the bottom of the river for him to slide under. The alternative was to keep crow-hopping into the deeper water until the boat pinned him against the cottonwoods and sucked him under. Gayeski reasoned that the raft was now the bigger threat to his safety. The only thing he could do was put the cottonwood between them, and the only way to do this was to go to the bottom of the river.

It was a questionable course of action, and one that could kill him just as easily. But there were no other options—and there was no time. He could feel the tugging of the current as the river, the boat, and the cottonwoods conspired to drown him.

Gayeski looked into the water and inhaled. He closed his eyes as he submerged himself, and the cold water took his breath away. Nearsighted anyway, he didn't need to see the underwater configuration as much as he needed to feel it. He forced his body to the bottom of the river, belly first and then chest, pressing himself down as he scrabbled along the gravel bottom, feeling only the vague shapes of rocks beneath his numb hands. He could feel there was space under the log, but it wasn't much. As he approached the opening, he flattened himself like a rat. Blind, with water gushing all around him, he squeezed himself under. The space was just big enough; the log brushed against his back as he wriggled to push himself through. The one thought he had was of his wife and children. If he made it through this, he would never leave them again.

Stanford understood what was happening just as Gayeski went under, and he started at a run for the cottonwoods. If Gayeski got caught, Stanford could attempt to cut him free, but the chances of success were slim. The supply raft had meanwhile lodged itself against

the cottonwood and tipped this way and that as the current pum-
meled it. Balanced on the raft's edge, the pinscher barked frantically
at the sudden disappearance of his rescuer, and was poised to leap in
after him. Stanford's eyes stayed locked on the water flowing out
from under the cottonwoods. The unimaginable had happened.
Gayeski had been washed through and deposited into the shallows.
He was alive and dripping, swaying as he caught his breath in the
ankle-deep water.

Shaking with adrenaline, Gayeski fought his way back upriver and
climbed over the cottonwoods to the supply raft. Stanford had
reached him by now and helped locate the bowline. Together they
tied the raft off to the tree. Stanford then ordered Gayeski back to the
gravel spit to get out of his wet gear. There was a serious danger of
hypothermia. Deeply cold, Gayeski peeled off his sodden clothes and
wrung them out, keeping only his waders on to break the wind. It
was 40 degrees out and past four o'clock. They were behind schedule.
The waders would have to do until they reached camp that evening.

The float team staggered off the river at seven, physically and emo-
tionally spent. Sergei the cook gave them a break from borscht and
the ubiquitous mystery-meat cutlets, and prepared salmon "Russian
style"—fried and covered with melted cheese and mayonnaise. The
men noticed with some unease that the pinscher had abandoned his
bear watch and refused to leave Gayeski's side. The dog spent the
night curled up outside Gayeski's tent, content to be in the vicinity of
his hero. The men spent a restless night waking not because of the
pinscher's alarms but because of their absence. Replacing the barking
was the sound of at least two bears circling their camp. Grunting,
rustling, and the snapping of twigs and branches filled the night's silence. Only the pinscher slept well.

The next day they took the morning to rest, fish, and collect data.
By this time, Stanford and Gayeski were just adding layers to what
they had already observed. It was clear that everything in the ecosystem benefited from the potent salmon fertilizer. That afternoon Stanford measured the biggest cottonwood he'd ever seen—fifteen feet
around. Guido collected samples of baby fish in little plastic bags,

where he could study their species markings. The only parr Guido had observed closely were the stranded minnows from Eagle Creek. Holding the bags to the light, Guido watched as the little fish quivered and flared their fins. They were perfect, beautiful tiny creatures, each with the miniature markings of its elders. Soon Guido was able to identify their differences by looking down at them in the river. The most exotic parr were those of the masu, or cherry salmon, which had deep bodies and exquisite leopard spots dappling their sides. The team broke for a lunch of brown bread and slabs of salami, cheese, and pickle, while Sergei the cook enjoyed his preferred meal of chunks of bacon fat with mayonnaise.

Back in the river, it wasn't long before Stanford detected another serious obstacle lay ahead. Rounding a bend, the ragged platoon of rafters rowed with renewed energy for shore. Stretched before them was a chasm where the river pushed between cliffs and an enormous boulder field, creating a long, tumultuous rapid. Rising from the white water were jagged car-sized rocks that could easily puncture their rafts. As they took in the gauntlet of rocks, a member of the expedition murmured to Guido, "I think we should call it a day. Let's radio for the helicopter." Guido said nothing. The news of the broken radio had not been shared with everyone. He finally responded simply, "We can't."

The only way forward was through the boulder field. Once again Stanford took over and instructed the men to climb to the tops of the most dangerous boulders and with their legs push the rafts away from the piercing rocks. Stanford would guide a test raft through the rapid to see it if was possible to maneuver the other rafts through. The men crawled out to their designated rocks and positioned themselves for Stanford's raft. When Stanford appeared at the top of the rapid, they managed to push his raft away from the sharp outcroppings as Stanford descended pinball fashion to the other end. When he shouted back that the rapid was doable, the men used ropes to guide the other rafts down, and one after another the rafts passed safely through the hellish stretch of water.

There were three logjams to negotiate in the afternoon, one that required the tedious exercise of unloading the rafts and carrying everything around onshore. The men worked as quickly as they could, forming a chain to unpack and transport the carefully arranged stacks of equipment, clearing brush to make a place for the coolers, tents, and boxes, and then lifting the boats over the jam with ropes.

The next jam had Sergei hacking away over a deep pool when the ax head fell off, plunked into the water, and was gone for good. The men paused in their exertions and stared after it in silence.

CHAPTER 14

TRUTH AND BEAUTY

DESPITE THE RELATIVE SAFETY and comfort at base camp, things were not going well. The fishing was mediocre, or worse. The steelhead were small and few. Some thought the run hadn't started yet. Others thought it was due to intense and illegal net fishing by poachers down by the river's mouth. Pete's anglers were frustrated. They had paid a considerable sum and traveled far to catch Russian steelhead. The salmon and char were plentiful, but Pete and his buddies had come for steelhead.

Pete tried the radio again. He had still heard nothing from Guido. As the days passed, he considered the possible scenarios. The expedition could be in trouble; someone might be injured; they could have lost a raft, or equipment. They could have had a bear incident. They had first aid kits, but a serious mauling would be more than bandages could treat.

If Guido's trip was going as badly as his, they would have to worry about future funding. Without good science or fishing, the Krutogorova might prove a total bust.

On the tenth day, the exploratory rafts rounded the last bend before base camp and Pete exhaled. He and Guido shook hands and exchanged brief reports. The research outcome of the float trip had been even better than expected. Stanford was nothing short of triumphant with the confirmation that an untamed salmon river was a bastion for all riparian life-forms. Guido was thinking they'd been lucky. Extremely lucky. He let himself feel the relief of having reached base camp with all his team members safe.

That night Stanford joined Savvaitova and Pavlov in their research tent to compare notes. While there was a dearth of steelhead at base camp, they found much to be excited about, and stayed up late into the night talking and drinking vodka while Savvaitova demonstrated splitting fish heads like a samurai and surgically removing the two tiny otolith bones that fit on the tip of her pinky finger. Stanford quickly learned that his Russian colleagues were at least as smart as he was—if not smarter in some areas. But, like their countries, they had a fundamental difference in orientation. The Russians had been focused mostly on the fish and Stanford mostly on the rivers. It was clear to both parties that they had something to teach each other.

Encouraged by Stanford's interest, Savvaitova recounted how her paper "The Noble Trouts of Kamchatka" had detailed her discovery of an unknown diversity within Kamchatka's rainbow trout and steelhead populations. She explained to Stanford that the fish she studied didn't conform to the two accepted life histories for the species. The story was hidden in the scales and otoliths of the fish, which she had learned to read like an oracle, accurately divining from microscopic markings, among other things, both the age of the fish and if the fish's parent was oceangoing. What the otolith bones of Kamchatka's steelhead told her was that they had at least six or seven variations on a life history. These steelhead didn't just go to the sea, return, spawn, and die. Some of them went twice. Some of them went six times. Some would go to the ocean for six or seven months and come back; some would go for three years and return. There were subpopulations enacting yet further combinations of salt-, estuary, and freshwater stays. It seemed that this natural population of fish hedged its bets, depending on the availability of food. Evolution had taught them that some years the ocean wasn't as good for food as the river. Some years it was the opposite. It seemed in steelhead, diversity was genetically coded in order to create a higher chance of survival in their complex and ever-changing ecosystem.

Savvaitova observed four major paths—a steelhead could stay in the river its entire life, it could travel downriver as far as the estuary, it could go to sea but remain close to the coast, or it could take to the high seas and travel untold distances. And then there were the anomalous "dwarf males" that never left the river. Smaller and less hardy than their rainbow trout brethren, these somewhat sneaky fish could spawn with all female forms, including steelhead. Low-risk and somewhat ignoble, dwarf males nonetheless kept the numbers stable in the unstable conditions in Kamchatka, where water levels and temperature varied greatly. Like the men who minded the home fires instead of going to war, dwarf males stayed behind in the river while their brothers risked the open sea. Savvaitova believed that dwarf males were a strategic reserve for the species, against whatever might happen to the larger males lost at sea.

It seemed that steelhead were capable of breaking the rules, and creating new ones. But why and how? Unlike their salmon cousins, who traveled through their lives with the predictability of a Swiss army, steelhead marched to their own drum. Perhaps steelhead had adopted adaptability itself as a genetic trait, and in doing so, challenged the assumption long ascribed to the natural world—fixed behavior. Evolution was thought to occur slowly, triggered by a freak maladaptation that, at a moment in time, proved a survival advantage—like the large ground finch of the Galapagos Islands, one of many finch species studied by Charles Darwin, notable for its anomalous short, thick beak strong enough to crack nuts; once a freak, this nut-cracking beak became the norm. Over the course of decades or centuries, the freak trait gained stability, and a new species evolved. But steelhead were doing something different. They seemed to change or alter their course within a single lifetime.

Stanford was fascinated. This adaptive plasticity would greatly increase their resilience as a species. Was it possible, he wondered, that the complexity of the river drove the diversity of fish? But the Russians hadn't yet grasped the concept of the third dimension. He would have to open their eyes.

The next day Stanford accompanied Savvaitova and Pavlov on their search for juvenile steelhead. The Russians had been scouring the river for days but had yet to find a single baby fish. They knew there had to be some somewhere. When Stanford saw the stretch of river they'd been searching, he suggested they take a walk in the other direction. "This water is too deep and fast for parr," he explained as they ventured out onto the floodplain.

The Russians were skeptical as they followed Stanford past camp into the tundra. At a confluence of clear tributaries, he stopped. "See there?" he said, pointing. The Russians looked down. There, in the floodplain of the Krutogorova, Stanford's mythical third dimension came to life, for swimming in the little streams and tributaries half a kilometer from camp were dozens of baby steelhead.

"That was the beginning," Stanford remembered. "From that moment on, they were on board. We realized it was possible that we had two halves of the picture—that together we might make an unprecedented contribution to salmon ecology." The next morning Stanford informed his team that he needed at least three men to help him walk the longest transect yet. Close to base camp, this transect would cross the wide floodplain near the mouth of the river. The news for the weary rafters was not uplifting. The only way to get across the dense forest and lush vegetation of the floodplain would be on bear trails. Stanford alone was energized. He had come halfway around the world to see these wide braided tributaries. In the back of his mind was the unlikely chance that he would see what Western scientists didn't think possible: all six species of salmon commingling in the floodplain.

After coffee and breakfast, Guido, Gayeski, and the Sergeis joined Stanford. Sergei the cook was armed with bear spray, the pinscher, and an AK-47. Stanford had chosen the location where they would start their trek across the river bottom, from one edge of the canyon to the other, a distance of almost two kilometers. Along the way they would measure and record vegetation, count and characterize the many river braids, and test the water chemistry and temperature to see if it was hospitable to all six species.

Fertile with dead salmon, the leafy growth was thick and grew far over their heads. Visibility was no more than a few yards in any direction. The five men walked in a line, Stanford in front and the cook at the end, facing backward with his AK-47. As they disappeared into the brush, the men called out continually to scare the bears, making as much noise as possible.

Progress was slow. Stanford stopped to map each creek and rivulet, recording its coordinates and altitude, all while the pinscher vibrated with intensity. The bears were nearby. The men could smell their dense, tangy aroma and hear the branches break as they lumbered off into the brush.

Stanford proceeded with calm deliberation, cautiously stepping around a steaming mound of bear scat, his hand raised in warning. Not long after, they came upon an interrupted meal of chum salmon, bit clean in half, its blood still running. The men called out more frequently as they crossed a clear channel marred by a single plume of just-disturbed mud.

The bears continued to spook as Stanford led the men across the floodplain, wondering how far in this widest possible transect the third dimension reached. A few hours in, the leafy vegetation gave way to forest, shot through with streams. Stanford and Gayeski dove their nets into the streams like kids, storing samples as they went. At noon, the men stopped to drink water and wolf down salami and cheese sandwiches. Stanford's eyes restlessly surveyed the reflective tributaries around them. In a few days they would pack up and he would make his way back to the University of Montana and spend the next months writing up his findings. He was still missing one crucial piece of evidence, however, and as they continued walking, his eyes scanned the water for signs of movement.

Two kilometers north of base camp, they entered the closest thing Stanford had seen to a stream utopia. As they sloshed through the verdant side channels, he recognized every possible type of stream: the crystal-clear spring creek; the tannic, tea-colored streams from forest runoff; the braids from the main channel, which rose and fell in elevation; the swampy, stagnant eutrophic streams, filled with algae. Reading the floodplain like a road map, Stanford called Guido over to show him how one stream dove under the river and popped up again in a high-nutrient "upwelling zone." He pointed out that chum salmon loved these zones; they were the right temperature, and the clean upwelling water prevented their fragile eggs from being smothered by mud or silt. Sure enough, in the next upwelling, they found some late-season chum spawning. The two men watched in silence as the fierce, snaggletoothed males monitored the roe-laden females, their bands of red, green, and yellow flashing as they scuttled the gravel with their tails to create a nest.

Soon after, they came to a young stand of cottonwoods and Stanford froze. The men behind him paused as he stared down at the water. Without taking his eyes from the stream, Stanford reached in his pack for his dip net and swiftly dunked the net into the water. There followed one of the most important moments of his career. He put his hand underneath his catch to make sure that nothing slipped out. Flashing and floundering against his palm were six species of juvenile salmon.

Stanford was speechless as Guido peered over his shoulder. "My god," he said.

"Take pictures, Guido," Stanford instructed quietly. "Quickly." He couldn't take his eyes from the net. The delicate parr were no more than four inches long. They had their scales and vaguely resembled the adults they would become.

The baby fish in his palm were all the proof Stanford needed. The interstitial flow of a floodplain river created multiple microhabitats— a mosaic of little channels that threaded and laced their way back and forth across the floodplain, reaching far and wide into the interior and deep into the forests—offering nearly every species of salmon the optimal conditions in which to spawn and rear and thrive.

Stanford's shock gave way to exultation. He had found fish Eden, the perfect salmonid habitat.

It was the second of two great discoveries, he thought, lowering the net back into the water and setting the little fish free. The first was that the abundant salmon of the Krutogorova turned the river into a biological force. The nutrient subsidies from dying salmon were directly responsible for the Krutogorova's strapping ecosystem, from the rich microorganisms to the oversized plants, bears, and eagles—all these life-forms had been nourished by nutrients from the sea. It was a magnificent interconnected system, and it was all held together by salmon.

With the Russians, there had been an unexpected third finding. Stanford believed that Savvaitova's discovery of steelhead diversity was as radical as his discovery of the river's third dimension. Salmon had survived centuries of habitat fluctuations in rivers and seas by evolving into species. Some years one species thrived while another declined; the next year it was the reverse. A healthy river supported this diversity. On a healthy river, there were always survivors. But Russian rivers were more than just healthy. Savvaitova had shown that on Russian rivers, evolution was still in play.

It was the hook they needed for both funders and the scientific community: salmon rivers were responsible for the greening in the Pacific Rim—for the region's biodiversity, productivity, and health. Because the salmon from so many Western rivers were vanishing, it was imperative that the salmon Eden of the Russian Far East be both protected and studied.

That night and for the remainder of the trip, Stanford, Savvaitova, and Pavlov bridged their cultural and scientific differences and forged a relationship that would continue for decades. The three scientists would establish findings they could only have advanced together, co-authoring over a dozen scientific papers. Savvaitova and Pavlov would go on to win Russia's top scientific prize for their discoveries on the Krutogorova.

"The truth is we'd been separated for far too long," Stanford said. "And together we made a hell of a team."

Guido watched the collaboration form with excitement, sensing the power of their combined science. After surviving a harrowing chopper ride over the Sredinny mountains, he gazed out at the rivers below and realized with regret that the Krutogorova would soon be laid with drills and pipelines. The UNDP initiative couldn't come soon enough.

CHAPTER 15

A Perfect Stronghold

In the autumn of 2000, Guido descended into Petropavlovsk through volcanic vapors with the glint of rivers below. The Moore Family Foundation had written the Wild Salmon Center a check for ninety thousand dollars, more than enough for Guido to play at the UNDP table. Meetings in Petropavlovsk would start in a few weeks, by which time Guido had to assess as many of Kamchatka's rivers as possible. If the Russians went for the stronghold approach, he had to be ready with a plan; he had to decide which rivers to fight for first. Because of its pipeline, the Krutogorova was not a contender.

It had been eight years since his first trip with Misha on the Bolshaya River. This would be a different kind of trip; there would be no leisurely floating down rivers, casting his fly from dawn to dusk, sleeping in the clean, cold air blanketed by stars.

He had been putting the pieces together in his mind. Xan's latest map had been a representation of the human population density of the Pacific Rim. The legend was a color spectrum that ran from yellow to red. Yellow indicated no people at all, orange indicated medium density, and dark red represented maximum clusters of 95,000

people. Kamchatka and the Russian mainland were almost completely yellow. Just inches below, Asia was another story. Japan and
the Koreas were mostly medium red, and next to them, China was
almost solid dark red. The country bordering Russia had a population of 1.4 billion, nearly one-fifth of the world's population. Because
they had run out of open space, the Chinese had built upward, erect-
ing skyscrapers that housed schools, hospitals, and supermarkets. As
it developed, China had given little thought to the state of its natural
resources. As a result, the land no longer yielded crops, and water
supplies were polluted. In some cities, the air was too toxic to breathe
without a mask. China had become one of America's biggest import-
ers of food because it could no longer feed itself.

Guido realized that with the global population doubling, it
wouldn't be long before food security became a serious concern for
the world's rising middle class, particularly in Asia. Perfect sources of
protein, like salmon, would be in increasing demand. The salmon
fisheries of the Pacific Rim were the biggest, healthiest protein factory
in the world; Kamchatka's salmon were feeding much of Asia. But the
source of this seemingly endless supply of salmon was in jeopardy.
Even the hardiest and most resilient fish on earth could not withstand
such extreme levels of poaching and overfishing. And now, with new
natural-gas reserves found off the western coast of Kamchatka, the
Russians were building a pipeline that would run from the north to
the south, across fifteen of the best steelhead and salmon rivers in the
world. The pipeline itself wasn't the problem. It was the roads that
would be built in order to construct the pipeline that posed a threat.
Roads meant access, and access meant poachers. The pipeline would
give poachers a superhighway with exit ramps to some of the richest
migration and spawning grounds in the Pacific Rim.

Measures had to be taken, and they had to be taken quickly. Guido
would need to convince the Russians that what they had in Kamchatka was extraordinary, that there was nothing like their salmon
habitat in the world. Second, he needed to impress upon them that
their moment of opportunity to protect Kamchatka would not last.
He had to make them see the unrivaled quality of their environment,

and he wasn't at all sure how to do it. He knew that to have a fighting chance, he would have to maintain distance between himself and the Cold War generals, who had all shown up for the meetings, eager to participate. He had spoken to Pete Soverel plainly about it. For the UNDP project to succeed, he needed to do things his way. This was not a time for power struggles or jockeying for control. Guido and Pete agreed that Pete would step aside; after all, the UNDP project had been Guido's idea. It was a careful piece of diplomacy that Guido had won, and he'd done it Pete-style, not by asking for control but by taking it. Now Guido was alone in unknown territory. His one wish was that the Russians hadn't already written him off as a spy.

As Guido made his way to his hotel, he noticed how Petropavlovsk had changed since he'd seen it last. Cruise ships from Asia were docked at the port; prostitutes, a sign of money to be made, lingered in the hotel lobbies, on the lookout for foreign clientele. Here and there the impoverished city flashed with conspicuous wealth: glitzy clothes, fur coats, and the occasional luxury car. These were signs of economic illness, of exploitation and a thriving black market fueled largely by salmon.

Guido spread out a detailed map of Kamchatka on his bed. There were only a few weeks to get up to speed, to gather the knowledge he needed to sell his plan to the Russians. He had to conduct an intensive survey of the rivers of Kamchatka before winter came and the salmon rivers froze. He needed a helicopter and a crew, and he needed a Russian site-selection team—local scientists, anthropologists, economists, and even lawyers who knew the region and the issues that went along with its protection. It was critical that the team was Russian and that the process was true to local interests. Guido had to watch and listen more closely than he ever had before in order to win over these weary and suspicious people.

In spite of the work that lay before him, Guido found himself preoccupied by Misha's most recent update on the rivers he'd been exploring. One report in particular triggered Guido's hunter's brain.

Misha, who was not given to hyperbole, wrote, "You're not going to believe this river." It was remote, even for the Russian Far East. Misha had stayed for days exploring, fishing, taking samples and notes. The Tugur was a big braided floodplain river that had seen few, if any, humans. Misha suspected it was a stronghold unlike any they had witnessed, for it held the rarest and most prized fish—Siberian taimen. At the mention of taimen, Guido wanted to know everything. Siberian taimen, the largest of all the taimen, were his new obsession, but unlike brown trout and chinook, these fish were a total mystery. Little to no data had been gathered because Siberian taimen lived in only a few rivers, and these lay at the end of the world.

Guido got Misha on the phone so he could grill him on the details. First off, had Misha managed to catch any? Yes, Misha said, a few. He had lost many more. The bad news was he didn't think it was possible to catch a mature taimen on a fly. Guido rejected this possibility out of hand. As far as he was concerned, every river-swimming fish could be caught on a fly.

Yes, Misha conceded, you might hook one, but landing it was another matter. In adulthood, taimen weren't just big, they were huge. Misha suspected it was because of their diet. He had observed them feeding on adult chum salmon, a twelve-pound fish. Taimen herded chum like sheep. And they weren't just fish eaters. They were said to eat birds and rodents too. But these fish were extremely hard to observe; they lurked in the deep waters of the wildest river Misha had ever seen. The taimen of the Tugur River were apex predators of an entirely new order.

But the Tugur and its taimen would have to wait. Here in Kamchatka, there was no time to waste. Before his eyes, the weather was turning.

In Kamchatka, the snow begins to fly in mid-October, though winter can always come sooner. It was mid-September when Guido was given a helicopter and crew, a site-selection team, and a map with many circles. He and a team of Russian scientists would try to see as many of these rivers as they could as they searched for the best salmon stronghold, a river that represented both the maximum biodiversity and bio-productivity—the most species and the greatest numbers of fish. Ideally it would be a river where scientists could set up field labs and begin to decode the mysteries of a perfect salmon habitat, a river without people or pipelines that could act as a natural laboratory so researchers could start establishing facts and findings that could be used to analyze, understand, and evaluate other rivers. In essence, they were looking for a Rosetta stone for salmon habitat. Anything less would not qualify for funding.

For the next three weeks, Guido sat with his face pressed to the window of their helicopter as the team flew low over plains, plateaus, and mountain ranges. The landscape was painted with the colors of autumn; the forests blazed with yellow, orange, and red. Near the rivers they had circled to visit, they landed where they could. The pilot carried a long wooden stick that they pushed into the earth from the open door while the chopper hovered, to test the firmness of the ground and confirm it could support its weight. Once they landed, the team clambered down the ladder and dispersed with their back-packs and notebooks to make their observations and evaluations. Guido did the same, but with a box of flies and a fishing rod.

Some of the Russians didn't understand what the American was doing, standing on the bank of the river, staring at the water in front of him—or why, after long still moments, he sprang into sudden mo-tion as he opened his rectangular plastic box and tied something to the end of his fishing line. What followed was something they had never seen before. Guido stepped into the river and waded deep into the current. Up to his hips in fast water, he lifted his graphite rod and moved it as if it were an extension of himself. As he cast the rod back and forth, the line grew long, and longer still, remaining under his complete control as it curled and laid out behind him. Then it paused, suspended, before Guido sent it shooting across the water to a spot in the river that only he could see. Without fully understanding, the Russians knew that what they were witnessing was mastery.

Those still skeptical of Guido's intentions considered that maybe the American wasn't a spy. It would be very hard to teach a spy to cast like this. Maybe he was what he said he was, a crazy fisherman who just wanted to save fish.

Guido and his Russian comrades saw some spectacular rivers: rivers that meandered through fields; rivers that were serpentine, bending rhythmically to the right and left; rivers that crashed down from the mountains in cold, clear torrents; rivers that braided four or five strands through the floodplains. At the mouths of some rivers were villages. Across the mouths of other rivers, like the Krutogorova, were gouges where the new pipeline was being laid. None of these rivers would be contestants for the natural laboratory.

One river held Guido's complete attention. It was wide and slow, flowing in big, lazy bends in an expansive loop through the tundra and mountains of the west coast of Kamchatka. Unlike the Krutogorova, it was a clear river, with an even wider braided floodplain. If there were fish here, this could be the most productive watershed Guido had seen. When they found a substantial gravel bar, the chopper put down. The vegetation was lush, and stemming out from the river were labyrinths of clear little streams. Guido grabbed his fishing gear and was gone. There were bear trails and the smell of the ocean on the wind. In the cottonwoods along the banks was the fluttering music of leaves. Guido stayed out for hours, climbing over logjams, inspecting rocks for aquatic insects, and pulling fish from the river, some of which he'd never seen before. There was a beautiful rainbow trout with white-tipped fins, and there were salmon and char of all kinds.

The group gathered for a lunch of cheese, tomatoes, and salami on the riverbank. Guido's gaze returned to the water as he considered the many species of fish he had seen. He wondered what he hadn't seen. Turning to the Russian scientists, he asked, "Tell me, does this river have chinook salmon?" With head nods, the Russians confirmed the presence of chinook. Guido proceeded, asking if pink and chum salmon also lived here, and coho. The answer to all three was *da*. The Russians added that the river also held cherry salmon. Absorbing this, Guido ventured another question. Did the river have steelhead?

Da, they answered. Guido went silent as they returned to their lunch. A few minutes later he hazarded a final question. There was one species he hadn't asked about, for sockeye salmon usually reared in the still water of ponds and lakes. They had found a few on the Krutogorova, but miles away from the main channel. The chance that this river supported sockeye was minimal, but he asked anyway. *Da!* came the answer. This river was so rich in habitat, the Russian ichthyologists maintained, that the sockeye didn't need a lake.

The river was called the Kol, and Guido's Russian comrades informed him it was a 540,000-acre watershed. Guido stayed long enough to know that the Kol held the greatest diversity of salmon, trout, and char he had ever seen—maybe in the world. It wasn't the prettiest river, but it was chock-full of fish. They flew over it from the headwaters to the mouth and found no human presence. It was, for Guido's purposes, the perfect first river to protect.

The next act unfolded in the drab offices of Petropavlovsk, where Guido faced the challenge of persuading the Russians to support something they did not yet believe in. They had tons of salmon; why did they need a program to protect them? Guido was asking them to accept a plan that, at a glance, defied logic. In a region filled with precious metals, minerals, natural gas, and oil—resources worth billions of dollars—they were to believe that salmon were potentially one of their most valuable possessions. Guido needed to make them see the big picture—that extraction of these other resources would destroy their land and rivers, and these resources would not last. The day the official meetings began, he buttoned his shirt, tied his tie, donned his tweed jacket, and willed himself into a state of optimism.

The building that housed the Ministry of Fisheries was an austere block of concrete with cold, barren stairwells and no working elevators. Seated in a small, sparsely furnished conference room was a complement of Russian district, regional, and federal officials. At the ready was a translator who stood awkwardly in the corner. Guido attempted to make eye contact with the Russians as they waited for the meeting to start and found nothing but wariness in their faces. He knew that if he failed to penetrate this barrier, he might as well go home. Speaking slowly so that the translator could keep pace, Guido explained that what the Russians had in Kamchatka was the finest salmon refuge in the world. The habitat on this peninsula was like no other, with rivers that hosted all six species of salmon, seven if you included steelhead. The Russians stared at him blankly: to them, this was a normal salmon river. Guido stressed that it was unusual—in America no river had all six species of Pacific salmon. Kamchatka's salmon rivers were the last perfect examples on earth, and they were in serious danger from development and poaching. Something had to be done to protect the fish and their rivers, or this enormous protein factory would be lost—to Kamchatka and the world. He was hoping that together they could defend this extraordinary habitat from destruction.

Guido's short, heartfelt presentation did not end with applause or even with nods of understanding. The wall of suspicion remained. Had they understood anything he had said? When the meeting ended, a kindly elderly bureaucrat approached him and bestowed him with an amused smile. He pulled Guido aside and pointed discreetly to a crack in the wall. "Remember you are in Russia," the old man said. As if to say, In Russia, someone is always listening.

There was one man at the table who had looked squarely at Guido throughout. He was younger with dark hair and a handsome, weathered face. His eyes, a piercing blue, seemed to take everything in. He was waiting for Guido outside the conference room and gave him a warm handshake. Vladimir Burkanov introduced himself as the director of Kamchat RYBVOD, the fishery-management agency of Kamchatka. Guido raised his eyebrows. This was the man who controlled what happened to everything that swam in the rivers and oceans of Kamchatka, and one of the most important people to the UN project. "I appreciate what you are trying to do," Burkanov said. "Tomorrow, come and see me at my office," he proposed. "We will talk."

CHAPTER 16

DEEPER CODES

G UIDO BREAKFASTED EARLY on bread and weak coffee and made his way back to the same drab building. Burkanov's office was up many flights of stairs, all the way at the top. It was an immense room with an equally immense map of Kamchatka hanging on the wall. Burkanov sat at a large desk with a telephone pressed to each ear. A third phone rang, unattended. He motioned for Guido to wait, he would only be a moment. When he had hung up all his phones, the Russian smiled. "Do you want to know what's going on here? Sit down, I will tell you."

To Guido's relief, Burkanov spoke passable English. He was a PhD in marine biology and had spent most of his career researching Steller's sea lions across the Pacific Rim. These benign and lumbering creatures shared the same migratory patterns as salmon and were part of the same ecosystem. Like so many creatures of the Pacific Rim, sea lions depended on salmon. Burkanov understood what it meant to lose salmon rivers to developers, and to lose salmon to poachers. He understood the trophic cascade that would ensue if Kamchatka's salmon population was decimated.

Burkanov smiled grimly, looking like something of a prisoner behind his desk. Guido recognized the signs of a trapped naturalist. The collapse of the Soviet Union had ended Burkanov's ocean research, and in the early nineties he was, like everyone, in need of work. His sterling reputation as a scientist and a man of integrity earned him a job offer he wouldn't have wished on his worst enemy. The task of overseeing and regulating Kamchatka's fisheries was in the hands of a corrupt agency of more than six hundred people, many of whom were fudging the lines or on the take. Along with patrolling rivers for poachers, the agency was responsible for policing the exclusive Russian Economic Zone, a one-hundred-mile belt off the coast that included large sections of the Bering Sea and the Pacific Ocean, as well as the Kuril Islands. But over half of all harvest was illegal, and it was the fish that were paying the price for the criminal activity and corruption. The salmon population was steadily declining, and Burkanov knew it.

Burkanov had given serious thought about whether to take the job of running the agency, knowing what it might cost him personally. He would be in control of an organization that had turned against itself. Bribes were a part of doing business, and fish were big business. In the new Russia, fish translated to food and money, and salmon of all species were routinely stolen, bartered, and trafficked. On Kamchatka's rivers, criminal poaching brigades laid waste to entire populations of migrating salmon. In the sea, the drift nets of four nations ensnared major populations of fish, snagging their gills with weighted nets often many miles long. Salmon were being cleared out of the oceans, and those that reached the rivers were not being allowed to spawn. It was a recipe for collapse. He was just one man, but in the end, Burkanov felt he had no choice but to do his best to prevent what was fast becoming a major environmental crisis.

Burkanov took the job. To patrol hundreds of rivers and thousands of miles of coastline, he had four helicopters, all in a state of disrepair, and four ships, none with fuel. It was impossible to police the waters or enforce the quotas of the Russian Economic Zone. Within the agency, corruption was the norm. Burkanov's inspectors confessed that they were forced to take the bribes and look the other way when they came upon fishing abuses, because they needed to put clothes on their children and food on the table. These were proud men, who had once wanted to do the right thing, Burkanov stressed, but their economic circumstances simply didn't allow for it.

Burkanov's phone rang. He answered it, hung up shortly, and continued talking. If Guido was going to start working in Kamchatka, he had to know what he was up against. With the corruption, it wasn't a question of right and wrong; it was a question of survival. Laws held no sway here. Russians weren't going to stop poaching and overfishing just because it was against the law. They were desperate. For them to change their behavior, their circumstances needed to change. Burkanov had reasoned that if enforcing the law became as profitable as ignoring it, things might turn around. There were considerable fines for poaching. What if he offered his inspectors a percentage of the fine for every poacher they apprehended? If he could create a countercurrent in the flow of cash, he might be able to save some fish.

He explained to Guido that he had drafted a new law for his boss in Moscow. For every fine garnered for poaching, 10 percent would be awarded to the arresting inspector. A new law meant little to the people of the Russian Far East, but word spread quickly that this one had a catch. It didn't take long for Burkanov to get his results. Collecting 10 percent of each fine added up, especially when there were so many violations. The flow of the salmon economy soon started shifting direction. In the next months, the poaching intervention rate, a previously nonexistent percentage, went through the roof. In the space of one year, the incidents of overfishing and poaching dropped by 70 percent. Kamchatka levied more fines than the rest of the country combined.

Burkanov requested that an additional portion of the fines be used to buy new helicopters and purchase fuel. As soon as they got their border ships in action, Burkanov's inspectors started cruising the high seas in search of criminal fishing boats. Newly energized, they conducted raids day and night, climbing over the hulls of offending vessels and making arrests. It could be a violent business. Many of these ships were real pirate ships, with fully armed crews. But Burkanov's men were armed too, and some confrontations ended with fatalities.

Burkanov's measures were working, but there was still a major gap in oversight. The ocean was wide, and it was easy for poachers to elude his men. He had read about new electronic tracking devices being used in New Zealand to monitor marine mammals. He had employed less sophisticated trackers with his sea lions, but these GPS trackers had microchips that transmitted data back to a network of computers so that the activities of the mammals could be observed in real time. On the computer one could pinpoint exactly where the creatures were in the ocean. Why, Burkanov reasoned, couldn't such devices be used to monitor the movements of fishing vessels to ensure that they were in legal waters, and to monitor what they were catching?

Soon Burkanov had a green light for a new tracking program, and Kamchatka had another new law. All boats fishing in the vicinity of the Russian Economic Zone were now required to carry a tracking device. In addition, the ship's captain was obliged to text a daily catch report detailing where they'd been, what kind of fish they'd caught, and how many. Burkanov and his men then analyzed the data and compared it to what they had on the computer and what they observed in their aerial surveys carried out by helicopter. The fishermen were now being watched from above, below, and within.

Not everyone was happy with Burkanov's innovative strategies. The Russian Mafia had long been invested in salmon poaching, both in rivers and in seas, and had many of Burkanov's employees on its payroll. Burkanov did not turn a blind eye to them, as his predecessor had done. When he uncovered their illegal dealings, he performed the revolutionary act of arresting them and sending them to jail. Some of them he knew personally, some were his friends, and they entreated him to help them, or at least let them off easy. Burkanov maintained he couldn't. This was the law, and it applied to everyone.

Burkanov had been well liked, but as his programs gained traction, his popularity went into precipitous decline. Even friends and colleagues distanced themselves, warning him that he should behave like a normal Russian or leave the country. If he didn't leave, they told him, he might be killed. Burkanov was too stubborn and principled to back down; there was too much work to be done, and, anyway, such threats were commonplace in the Russian Far East. He would know when it was time to get out.

One day Burkanov got a call from the director of the fisheries agency, who ordered him to suspend the investigation of an apprehended poaching vessel belonging to a government insider. Burkanov maintained that he couldn't stop the investigation; it was already under way. That night he was roused from bed by a phone call after midnight. Burkanov was told that the captain of the vessel knew where he lived. Other officers had given him Burkanov's address, and the names of his children. "If you don't release the vessel, if you continue with your investigation, your children will be dead by morning."

Burkanov didn't waver. He responded calmly that he would continue with the investigation and do his best to protect his children. This allegiance to a basic ethical code was incomprehensible to his peers. In Russia, such integrity was not admired; rather it was considered a personality disorder. Burkanov was advised again and again that if he wanted to survive, he should play the corruption game like everyone else. The more successful his agency was, the more isolated and vulnerable he became, and the more intense and personal the threats. Burkanov understood that they were not idle threats.

"Do you know the going rate for a human life in the Russian Far East?" he asked Guido. "Ten thousand dollars. For this amount, you can have anyone killed." But Burkanov couldn't back off; his measures were working, fish stocks in Kamchatka were rising, the illegal activity in the seas was a fraction of what it had been, and hundreds of thousands of fish were able to spawn.

Guido was beginning to grasp that the problem with the fisheries was the problem with Russia, and it was intractably deep, a cancer of corruption in the system that no one man could eradicate. He headed back to his hotel as comforted by his new colleague as he was unsettled by the immensity of the challenge and by the certainty that Burkanov was in no small amount of danger. Guido wondered if he had known anyone as principled or courageous, or if he'd ever had such a valuable ally.

The next day, Guido met with the Russian groups that would be supporting the goals of the UNDP project. These were the people who would coordinate with fisheries, research laboratories, and both local and native communities, working to collect data and gain compliance. These inglorious middlemen were critical to the success of the project. In the spacious conference room, their representatives were arranged in a circle. There was a lot of coughing and sniffling, and someone whispered to Guido that influenza had swept the city; even the schools were closed. One by one, representatives listlessly described the steps they would take to help move the salmon initiative forward. Guido didn't sense much conviction. If these groups didn't have a true sense of purpose, their commitment would only go so far.

At the end of the day, he thanked the Russian representatives for their presentations. Then, in an appeal to their national pride, he told them that when the salmon plan was implemented, not only would they be protecting invaluable habitat, but Russia would be the international leader in salmon refuge designations. It would be the first country to protect its precious resource, and its example would be recognized throughout the world. When the translator had finished, Guido looked around the table, and for the first time he saw glimpses of warmth and approval.

That night he called Burkanov to give him the update. The phone line connected briefly and then dropped. When Guido tried again, he heard the telltale click of electronic eavesdropping equipment for the split second before the connection was made. Although he had nothing to hide, he was spooked and hung up. He would see Burkanov at the meeting the next day when they would be hearing from the public stakeholders. Among them were scientists, native peoples, environmentalists, and fishermen. There was a host of issues to contend with, but there was only one that could derail the whole process.

Petropavlovsk often went without electricity for days and even weeks. One winter, a nuclear submarine powered the capital city throughout the day and night, thawing the bitter cold one quadrant at a time. The peninsula was desperate for energy. The new governor's act of mercy was to deliver the people from the cold and hunger of the long winter months by building a natural-gas pipeline. At present, the pipeline would also cross the mouths of fifteen of the most species-rich salmon rivers on earth, exposing them to environmental degradation and poaching. Guido's recommendation was to build the pipeline through the interior, avoiding the coastline and the rivers. While the pipeline plans had been made, they were still in the early stages. Plans could change.

The UNDP had a further objection: such a pipeline would provide gas for only ten to twenty years. The pipeline would soon outlive its usefulness. The bitter irony was that the project was being paid for with money from Japanese fishing permits to gillnet Russian salmon.

It was a scruffy man from St. Petersburg's Academy of Sciences who opened the discussion, and he did it with a simple question. How could such a good idea as the salmon project be followed by such a bad idea as the gas pipeline, which would be an environmental disaster for Kamchatka? What was the UNDP's position on this?

The UNDP team gathered in a quiet huddle. Any direct opposition by them would get back to the governor, who had made clear his reservations about the UNDP salmon program. The governor was the one person who could shut the initiative down. As his colleagues conferred, Guido suggested that it wasn't their job to interfere with Russian politics or become embroiled in the weighty issues facing the people of Kamchatka. They were not there to legislate what the people of Kamchatka should do. Maybe, he suggested, they should simply say that the pipeline was a Russian issue, and leave it at that. It was an elegant solution, one the UNDP team adopted.

The next question was directed squarely at Guido. It came from a slim, balding man representing the association of the native Itelmen people. He contested the supposedly humane practice of catch-and-release. "We have fifteen thousand years of experience with salmon," he began. "Our experience tells us that catch-and-release is fatal to the fish. It is also cruel because it damages the salmon's aura. It is like being pulled underwater and then being allowed back to the surface. We kill the fish very quickly, and only after apologizing. Therefore we object to this treatment of salmon and steelhead by the visiting tourists."

Was this true? Did this Western practice kill the salmon? Here Guido was backed by Moscow's premier ichthyologist, Dmitri Pavlov, when he described the numerous site studies that showed how catch-and-release did not harm the fish. The Itelmen remained unconvinced. Pavlov was a glaring representative of the Soviet system, and of Moscow. Moscow always had its own agenda; why should its scientists be trusted to decide local matters?

The group convened for lunch and seemed to return in better spirits. To Guido's relief, Burkanov offered to take over the meeting and proceeded to guide the discussion smoothly through the frayed patchwork of Russian civil society. By day's end he had persuaded the groups to agree on and finalize some key points. It would take many such meetings before the Russians gained a comfort level with these radical ideas about salmon. Thankfully, Burkanov stayed with the discussion every step of the way and, surely and steadily, marched the process forward.

Before he left, Guido met with his friend one last time. He was worried about Burkanov's safety. Burkanov admitted he was feeling the stress. He had developed ulcers and badly wanted to return to research on the open seas, but his work in Kamchatka wasn't done—nor was Guido's for that matter. The UNDP salmon project had yet to gain full approval; they still had to convince the governor. Guido had been thinking about this critical final step, and he assured Burkanov that when this meeting came around in a few months, he would be fully prepared.

CHAPTER 17

THE RIGHT RIVER

IT WAS FOUR MONTHS LATER, in early 2001, when Guido returned to Kamchatka. He sat with his face turned to the window as his plane dropped through the clouds, the smell of sulfur filling the cabin as the old 737 flew near the plume of an active volcano. On board was a crew of burly oil riggers on their way to the Sakhalin oil fields, an aid project manager, and a handful of Russians loaded down with American merchandise. Guido's bulkiest item was a slide projector and carousel he had filled with twenty-five carefully curated images, photographs he had taken over the years. He had given much thought to how he could best present his case to the Russians, and decided that pictures were the most efficient way to show them what they needed to see—the future.

It was winter in Kamchatka and the city was gray and cold. He and Burkanov met for a drink. The two men had kept in touch in the interim and were happy to see each other. Burkanov looked worn-out, and Guido wondered how long he could keep up his vigilant stance at the fisheries agency.

Burkanov sent his right-hand man to Guido's first meeting, since he could not attend himself. Guido had set up a screen, and when the room was full, he dimmed the lights and led the Russians through a procession of images. He started with arresting aerial shots of Kamchatka's rivers, showing them Kamchatka as they'd never seen it before. There were vivid photos of snowcapped mountains, rivers, and tundra. There were photos of flora and fauna, many of them fish. The Kamchatka Peninsula, he told them, was an ecological paradise. Their home was also home to all the salmon species in the Pacific. It was the only remaining model of what salmon rivers should be. In the darkened room, there were smiles and nods; their home was beautiful indeed.

As Guido clicked through more slides, he explained that there were no other rivers like this in the world. He then clicked to an image of the Clackamas River in Oregon, and paused while the Russians stared at it. It looked nothing like rivers they knew. It was hemmed in by agriculture and had only one meager channel that flowed, strangled, through fields and developments. Where were the spawning grounds? Where were the side channels where the juvenile fish reared? they asked. "Gone," Guido said. Was this what an American salmon river looked like? It's what they look like now, Guido answered. "Because the salmon are gone."

He pointed to the swaths of clear-cut forests that ran from the tops of hills all the way down to the river. Removing trees from a salmon river was very harmful for the salmon, he explained. The fish needed trees along their riverbanks to shade the water and keep it cool, and to create nooks and crannies with their fallen branches and limbs. The more of these little habitats in a salmon river, the better. The next slide was a patchwork of fields. "You see how close these fields are to the water?" Guido asked. "The farmers divert water from the river, and then drain their fertilizer and pesticides back in. There isn't enough oxygen in such polluted water for a fish to breathe." He clicked to a slide of the imposing Bonneville Dam, with its furious turbines churning thousands of cubic tons of water. How were fish supposed to survive this? Some of them made it up the fish ladders, but many more perished, battering themselves against the torrential falls, unable to reach their home rivers to spawn.

In America, wild salmon were disappearing. The American solution to this problem was to build hatcheries, Guido explained, but this only created more problems. Hatchery fish were not as adaptable or resilient as the wild fish, and they were outcompeting the wild fish for food and resources. Soon these weaker hatchery fish could replace the robust fish of the wild.

"So you see, this is how we lost our salmon," Guido summarized. "Every one of these things could happen in Kamchatka."

The Russians went quiet as they absorbed this grim warning about what might become of their exemplary watersheds. Guido clicked to an image of the Opala River, which came with the relief of a tall glass of cool water on a smoldering day. "Look at this beautiful river. You haven't yet allowed people to ruin it. You still have a chance to protect it, to get it right. I am offering a picture of our disaster, so you can learn from our mistakes." He followed this slide with images of road building and poaching around the Kamchatka Peninsula. "You see, it has already begun. I assure you, it can happen very, very fast."

Guido clicked on Xan's map of the Pacific Rim that showed how salmon were disappearing from its southern edges and explained that this decline was moving steadily north. Kamchatka was one of the few places left that was still unadulterated, still perfect. "The key is to protect these salmon rivers before the damage is done. And in the meantime, we can move a few of the best rivers into the protected park system," he told them.

The Russians talked among themselves. They had a question. How many people work in fish restoration in the United States, they wanted to know. Guido answered that thousands of people worked in fish conservation, which impressed them. These were real jobs. Maybe they could create such jobs in Kamchatka.

Guido's last slide was an ominous one. It was a map of the Sea of Okhotsk that identified areas for offshore oil fields and deposits of natural gas. "Russia is now at a crossroads," Guido said. "Change is coming; things you can't see. And once this development starts, you can't stop it."

The lights went up and Guido looked around the room, but the Russians were focused on the last slide, still faint on the screen. The room had gone silent again. One of the officials raised his hand and asked, "Where did you get this map?" Guido was caught off guard; he couldn't remember where he had gotten it.

After the meeting, Burkanov's associate pulled Guido aside. "You said you don't remember where you got the map, and that you don't know where this information came from. The problem is that some of the members of the committee think this map and its information came from the CIA." Despite all of Guido's work to dispel mistrust, he had landed smack in the middle of new and justifiable suspicion. He told Burkanov's aide that the map had probably come from the World Wildlife Fund. "What can I do to convince them it didn't come from the CIA?" he asked. This earned him a conspiratorial smile from Burkanov's man. "Don't worry, we will convince them together."

When Guido and Burkanov met again, Burkanov described how it worked in Russia. Companies and agencies like his were assigned "curators" from the FSB. These representatives from the secret police visited regularly, sometimes a few times a month, sometimes a few times a day if something serious was happening. The agent would evaluate the organization's goings-on and make recommendations. In the case of the Wild Salmon Center and the UNDP setting up shop, the curator for the Kamchatka fisheries was making regular visits. He had daily updates from his subordinates, who received reports on every meeting. The FSB was highly interested in the Wild Salmon Center and had been for years. They had had many arguments about what these Westerners were up to in Kamchatka. The fisheries curator had finally made a visit to Burkanov's office to have a straightforward talk about it. "Vladimir," he said, "we have been having many disagreements in our department about what is going on with this Wild Salmon Center. We think that what is happening is that they want our wild salmon to go to the United States. They have no more salmon in the Northwest. They want to get our Russian salmon to swim to their side of the Pacific Rim and to swim up the Columbia River so they can catch them."

Burkanov looked at the man and considered. "And how would they do this?" he asked.

"There might be some chemical that changes the genetics of the Kamchatka salmon. It will make them swim to the Columbia River."

"Okay," Burkanov said, keeping a straight face. "That's an important issue to address."

"Many people in our department say it's possible," the curator continued, "that they can do such things. We think this is the reason they have come here. Why else would they care so much about our salmon?"

"What about the other people in the department?"

"Other people think it's impossible."

"What do you think personally?"

"I think it's impossible. But I want your advice. Is such a thing possible?"

"It's a very good question," Burkanov agreed. "But no, it's not possible. Even if it were possible, do you know what these Kamchatka salmon meet first in the Columbia River? Dams. Not just one, but sometimes up to five."

"What's a dam?" the curator asked.

"Hydroelectric plants that block the river. They give no place for spawning, and no place to get past the dam."

This was news to the curator.

"Yes, and they have pollution there, and they've destroyed all the natural habitats of salmon, and now their salmon have no home. That's the reason they support protecting our salmon. They're trying to teach us not to do stupid things like they did."

"I see. You've made a good argument. If it's nothing but dams over there, there's no sense in changing the genes of salmon."

"Exactly," Burkanov replied. "Kamchatka salmon would swim all the way to America, and when they got there, they'd see all the pollution and the dams and they'd turn around and come straight back to Russia."

The curator went off, satisfied with the answer.

Guido laughed along with Burkanov as he recounted this conversation, though it was a sad fact that Russian salmon would never choose American rivers over Kamchatka.

Burkanov assured Guido that he would be there for the meeting with the governor. The next day, the UNDP team was there waiting for Kamchatka's most powerful man, prepared to offer their best diplomatic efforts. This much they knew: the governor was willing to discuss the salmon project, but there was no chance that he would back off the construction of the pipeline, despite the vehement opposition of local environmentalists. As the governor delivered his speech, extolling the pipeline's environmental safeguards and the benefits to Petropavlovsk, he watched Guido out of the corner of his eye.

The UNDP team was also watching Guido. They feared that as the head of the Wild Salmon Center, Guido might vocalize his opposition to this direct threat to one of the last salmon strongholds on earth. They advised him to do nothing to endanger their relationship with the governor. Guido listened and nodded, but he had already come up with his own strategy. He knew by now how desperately the region needed the influx of money and resources the UNDP would provide, but not at the expense of heat and electricity. If the people of Kamchatka didn't have energy, they didn't have a functioning society. As Burkanov had shown him, you couldn't make someone choose between basic survival needs and the environment.

At the end of the governor's address, Guido was invited to comment. He nodded politely to the governor and rose, directing his words to the translator. "Thank you, Governor. We understand your need for a pipeline. For the part of the Wild Salmon Center, I would like to add only that we would like to offer any environmental technology to help reduce the environmental impact of this development."

The governor brightened at this unexpected clemency, and then slapped the table with his hand and said, "We do not just request this information, we demand it!" Guido took the opportunity to suggest that they protect their rivers by policing the roads along the pipeline for poachers. The governor agreed to consider this, and in time he followed through by designating pipeline roads as private and adding guard stations.

The afternoon was momentous. The Russians voted to support the stronghold strategy, agreeing to move one river into the protected parks system. At the end of the long day, they still had to choose a first river to protect. Guido reminded them that the program would work only if they chose the right river. He could not dictate this decision, he told them, though he knew what that river was.

What followed was an extended discussion between the Russians about which river to choose. Guido listened to much opining as they shared their feelings about the various virtues of their favorite rivers. Guido reminded them that they would have to choose a river that contained all the Pacific salmon species—that this would represent Kamchatka's ultimate stronghold. He added that without this condition, they couldn't get the co-funding for the project. After conferring some more, the Russians made their decision. "We nominate the Varavskaya River!" they said.

Guido had flown over this beautiful river with its forests and waterfalls. He had seen its clear water and majestic cliffs. He had also seen the large village at the mouth of the river. With so many people living there, he told them regretfully, the Varavskaya could not qualify as a natural laboratory.

The Russians grumbled and returned to their discussion. Finally, from the end of the table, Burkanov, who was privy to the site selections Guido had made, suggested the Kol River. Guido nodded a brief thanks while the suggestion met with immediate dissent from the others. "The Kol's no good!" On this, the Russians agreed. It was not aesthetically beautiful—it had no splendid cliffs or waterfalls, and it was filled with logjams. Why should this dull, unattractive river be the one they protected?

Summer afternoon at the Deschutes, Lee Rahr,
Tucker Malarkey, and baby Henrik.

G. RAHR

Gee, age six, with bull snake.

G. RAHR

Poachers' camp and piles of pink and chum salmon killed for their roe, Sakhalin Island.

DMITRY LISITSYN

Huge bag of salmon roe confiscated from poachers' camp, Sakhalin Island.

DMITRY LISITSYN

Approaching the Tugur River, autumn 2015.

G. RAHR

Alexander Abramov fishing a tributary of the Konin River, Tugur watershed, 2015.

G. RAHR

Our camp on the Tugur River, first night, 2015.

T. MALARKEY

Ivan, our Nanai fishing guide,
Tugur expedition, 2015.
T. MALARKEY

Guido and Tucker.

JOHN JUDY

Guido and Misha Skopets fishing on a cold day on the Tugur, 2015.

T. MALARKEY

Guido and eighty-three–pound Siberian taimen caught on the Tugur River, September 2018.

MATT SLOAT

Brown bear and cub eating salmon, Kurile Lake.

IGOR SHPILENOK

Brown bear sniffing the air, Kurile Lake, Kamchatka.

IGOR SHPILENOK

Guido spey casting on a coastal salmon river, Oregon.

JEFF STREICH

Guido explained why the Kol might be the perfect river to protect. On the Kol there was no hatchery, no streamside road, and no tragedy of declining populations. The Kol had superb habitat, with abundant runs of wild salmon, including all six species: chinook, sockeye, chum, coho, pink, and masu. It also held rainbow trout, steelhead, and two species of char. In the previous year, over four million fish had returned to spawn, filling the river so fully that in some stretches salmon were packed side by side like paving bricks. Guido concluded that while the Kol might not be as beautiful as other rivers, it had the highest salmon diversity known in Kamchatka and would make a fine first stronghold.

The next day, Guido and Burkanov tag-teamed, and by lunch the federal ministers in the room began to concede that the Kol might be the river to protect. When they reconvened after lunch, the local officials joined in and warmed to the idea. By afternoon, they unanimously agreed that the Kol would be the best first river.

"There will be other rivers to protect," Guido assured them, "some of them very beautiful." He took the opportunity to write out the names of four additional rivers that he and the Russian scientists had thought about carefully. The five rivers represented the full spectrum of species diversity in Kamchatka, and Guido and his colleagues wanted them all protected eventually. It was a risk to push this bigger agenda now, but to get these rivers on the map was critical, and he didn't know when he'd get another chance.

That afternoon Guido and the UNDP team watched as the governor signed the official agreement to support the protected-areas project with the UNDP. It was a huge triumph, and Guido left the meeting feeling ebullient. They had secured an agreement to protect a 540,000-acre watershed with the greatest diversity of salmon, trout, and char in the world—and it would be protected, the whole watershed, all the way to the ocean. Four more rivers had been advanced and were on the map for consideration. It was Livingston Stone's dream, and it was happening in Russia.

Flying home, Guido was surrounded by international oil drillers and gold miners animatedly talking about their latest deal with the cash-hungry Russians. He looked out the window. The trees and forests were still in their autumn colors, but the hills and mountains had been softened by a blanket of fresh snow. Seemingly endless ridges of mountains spread to the north, with the broad Kamchatka River valley visible beyond. From the center of the Kamchatka valley, the Klyuchevskoy Volcano rose more than 15,000 feet. Below was a broad lake, covered with ice and surrounded by a lunar landscape of volcanic craters and three majestic volcanoes. And beyond that, to the west, was the bountiful Kol, flowing strong to the sea.

It was a few months later that Guido heard that Burkanov's adversaries had ludicrously accused him of corruption and had him targeted by the FSB. Local newspapers had started to smear him. He told Guido on the phone that he and his family were finally considering leaving Russia. There had been too many threats; his family could no longer lead a normal life. "It's like trying to fight with the wind," Burkanov said.

Russia had no use for Burkanov, but America certainly did. In fact Guido had been considering asking him to join his board of directors. The next phase for the Wild Salmon Center would be focused on deeper scientific research, and further unlocking the secrets to the productivity of Kamchatka's rivers. Who better suited to advise him than a premier scientist like Burkanov? As Steve Beissinger had argued at Yale, science was the key weapon in conservation. In Guido's stronghold war, the most valuable generals were proving to be people like Xan and Burkanov. Without them, Guido was, as Beissinger put it, just another citizen with an opinion. In Russia, science had been of utmost importance; it had given the Wild Salmon Center a respected and safe arena in which to operate.

Burkanov told Guido that the NOAA, the National Oceanic and Atmospheric Administration, was interested in hiring Burkanov at its Seattle office to continue his work on sea lions. He would be able to return to ocean research, where he belonged. And in the meantime, he was happy to join Guido's board and agreed to fly over and visit the offices, meet the Wild Salmon Center team, and attend a few meetings. Guido had one meeting in particular in mind. For the next round in Kamchatka, they'd need funding to set up research stations and mobile laboratories. Then they could begin the all-important collection of data. This was how policy shifted. Guido had started to think that Burkanov would be the ideal person to bring to meet the man who could change the fate of both Russian and American salmon rivers, a brilliant science geek from Silicon Valley looking to invest in the environment. Guido thought the two of them together had a chance of convincing Gordon Moore that Russia was worth his further consideration.

Two months later, Burkanov was an official director of the Wild Salmon Center's board, and accepted Guido's invitation to come to Portland. Guido introduced the valuable new board member to the staff, and showed Burkanov what they were working on. Burkanov took particular interest in the watershed councils that Governor Kitzhaber was setting up along Oregon's beleaguered rivers. Local stewardship groups struck Burkanov as an effective approach to river protection, and one that might work in Russia.

A few days later, they boarded the quick flight to San Francisco and drove down the peninsula to Palo Alto, where the Moore Foundation was nestled among the other Silicon Valley giants on fabled Page Mill Road, with its manicured grounds and sprawling compounds glittering in the California sun. Guido wasn't worried about meeting Moore. That Moore was a scientist and a fisherman meant Guido could talk to him plainly.

They didn't find Moore in a luxurious corner office overlooking the forested hills; he was seated in a cubicle, alongside everyone else. When he rose to greet them, he looked them up and down and smiled. Guido noticed that his eyes rested on their shoes, which were brown, dusty, and scuffed. Guido and Burkanov had dressed up, but shiny shoes were not in their wardrobes. Moore seemed reassured by this detail that spoke to equal parts dedication and desperation.

Moore had read the preliminary report on the Krutogorova. Now he listened as Burkanov laid out the situation in Russia, describing the broken system, the poverty of the people, and the conservation crisis with the salmon. Guido added that they were in a race against development. The situation was dire, but the Wild Salmon Center was well positioned, with the UNDP projects up and running, scientific allies in Moscow, and local support in Kamchatka. If they acted now they could fend off these threats with scientific research and findings. The Krutogorova was just the beginning.

Moore saw no problem with the plan. The question was, why would an American billionaire sink money into a resource-rich country already full of billionaires? The answer, Guido explained, was science—the Russian Far East was a big laboratory. It was vast and deep and promised discoveries they could never make in America, as well as insights that could guide them in their own conservation. For the first and maybe the only time, the door was open to them to experiment and observe. It was a moment they had to grab.

Moore agreed but insisted there had to be some degree of measurability—a way to evaluate the effects of his investment. Moore was a scientist, and this was the essence of science. With nature you also needed feedback. To understand the impact that conservation efforts had on an ecosystem, one had to be able to assess outcomes—an increased number of birds, animals, and plants, for example. Burkanov concurred, and shared his own experiences with measurability, first with researching sea lions and more recently with the trackers he had slapped on fishing vessels in the waters around Kamchatka to keep them honest. But how to do this with salmon? Unlike sea lions or ships, they lived entirely below the surface. There were millions of them; you couldn't possibly tag enough of them to get a complete picture.

Guido insisted that salmon were measurable. They passed the same point twice, on their way to and from the ocean. They could be counted, and they could be caught. Their otolith bones and scales could be analyzed. Salmon were regular messengers; their messages only needed decoding. Research stations and mobile laboratories would make this possible. They now had access to more salmon rivers than they had ever seen before. Perfect, untouched rivers and fish. What kind of messages would they carry? The Moore family's support could help them find out. Moore was intrigued. He told them he would continue his support, and would be interested in hearing what came of it.

CHAPTER 18

BIG FISH

IN THE OFFICES OF the Wild Salmon Center, Xan was getting close to completing her data-gathering on the distribution of salmon across the Pacific Rim. She had formed agreements with the fisheries agencies of Alaska, Canada, Washington State, and Oregon, and they had started handing over their information on the status and health of their fish populations. Russia came last; salmon status there was a national, state-controlled secret, and Xan's repeated requests for data on Russian salmon runs and distribution had gone unanswered. Burkanov, who personally knew the people who gathered the data, vouched for the Wild Salmon Center and assured them that Xan was a legitimate Western scientist, not a spy. With Burkanov's help, Xan was finally able to open the black box of salmon status in the Russian Far East.

This was just the beginning of building out the map. The next piece was to pinpoint the location of fish hatcheries and find out how many fish and what species these hatcheries were producing. Guido's plan also included an assessment of the condition of the watersheds and salmon runs from Oregon to Japan. The data was proliferating. In fact they were no longer calling it a map; they were calling it an atlas.

While Xan built her atlas, Guido focused on winning protection for rivers in Kamchatka. Kamchatka's governing bodies now fully acknowledged the wisdom in protecting salmon, declaring them the foundation of the region's economy. Government officials went on the record stating that the temporary riches yielded by mining, oil, and gas developments could not compare with the gift of salmon. If they could sustain salmon, salmon would sustain them. Salmon provided food, revenue, cultural value, and scientific investment. Kamchatka had proudly recognized itself as home to the purest genetic reserve of the world's most valuable fish, and had committed to becoming its protector.

The conservation wins were amounting to something. The Kamchatka government advanced programs for antipoaching, environmental education, and supporting indigenous groups in the defense of their home rivers. The Kol River basin would be the world's first whole river salmon refuge, equaling 540,000 acres. In northern Kamchatka, the Koryak consented to list the Utkholok and the Kvachina basins for future protection. The Oblikovna, Zhupanova, and Bolshaya rivers would be the focus of science and conservation programs. Under the plans, these watersheds would be protected from development and two would be given biological stations. It was a list made by Guido and the Russian scientists, but the Kamchatkans had made the stronghold approach their own. In Moscow, Savvaitova and Pavlov were bringing in other Russian scientists and fish managers and actively creating their own Russian stronghold map.

Burkanov had moved to Seattle, but before he'd left Russia, he'd done his best to fill Kamchatka's fisheries agency with worthy protégés, people who understood the work and knew its value. Guido and Burkanov, who had benefited so greatly from their own alliance, saw the advantage of setting up more such partnerships, and formed an environmental exchange program that would draw a steady stream of Russians to the Pacific Northwest and Alaska, where they could see for themselves how the other side of the salmon watershed worked. In the next years, Russian fish managers and students came to Alaska, Oregon, and Washington State, where they were openly dismayed by the dams and hatcheries but were inspired by the watershed councils. As Burkanov had predicted, this was something they felt they could take back to Russia.

Creating a salmon atlas wasn't Xan's only job. She had become indispensable as Guido's number two. Xan, with her brilliant and agile mind, seemed to have bottomless capacity. Unlike Guido, she could handle both human and scientific systems, deftly sorting through the administrative details that Guido preferred to avoid. With her even temperament and reliability, she was the engineer that kept the gears of the Wild Salmon Center oiled and turning. She and Guido were a team, and together they had grown their organization. There were thirty people on staff now. While Guido could inspire and direct his new employees, he had little time to manage them. Xan unhappily found herself forced to take on the role of the heavy as Guido flew ahead, tethering him to the ground when necessary, and acting as a buffer to the many employees who were bewildered by him.

Guido was often too occupied with his own program to notice his shortcomings as a manager. He was expanding the Wild Salmon Center to Washington State, the Oregon coast (where he was fighting timber companies on the Tillamook), and Japan. In his mind was the sound of a ticking clock. Across the Pacific Rim more and more hatcheries were being built, and the pressure to develop the Russian Far East was mounting. Since Burkanov had left, incidents of poaching and overfishing had exploded. Even in protected areas, there were those who doubted that it was possible to enforce protection against such a powerful criminal element. From where Guido stood, Pacific salmon were still on track to repeat the scenario of Atlantic salmon, who had disappeared from their once immense habitat and now could be found only in the northernmost rivers.

Guido was struggling with another ecosystem, one that he himself had created. He was a husband and a father now, roles that he appreciated on a biological level and did not want to fail at. When he was home, he devoted his attention to his towheaded son, who was showing the same curiosities about the natural world his father had shown. But he wasn't always home. Lee did her best to be patient, but she was frustrated with the amount of time Guido spent away. Still, she could see he was doing what he'd set out to do, and she did what she could to support him. She kept house, played hostess to the Russian scientists when they came to visit, and occasionally caught glimpses of the complex world Guido was inhabiting. Most of the time, though, he was a blur to her. So many things were happening with him, she could barely keep up. They made sure to spend weekends at the Deschutes, where time slowed and they were able to sit across the table from each other and see each other clearly. Their place in the canyon prevented signals of any kind from reaching them, and for a few days at a time they were spared the beeping and ringing of their increasingly technological life. They lived like they had in the beginning, taking walks, cooking dinners, and fishing—now with little Guido in the backpack. The sound of the river soothed them as it always had, delivering a peace they could not find at home.

I noticed that Guido was making other adaptations. After years of watching, he was learning how to deal with people. While profoundly confident in his knowledge of the natural world, he had never taken for granted that he understood humans. He approached them with humility and openness, relying on his observational skills and instincts to assess their character. Over the years, he'd seen that people, just like nature, were governed by certain laws. That he was capable of discerning these laws made them infinitely more interesting, and it made them useful. If he could sway them to his cause, there were no limits to what he'd learn. His social skills had advanced; he'd discovered how to draw people out and actively listen. Combined with his other talents, this led to a move up the food chain, where he was able to cultivate relationships with people who could help him on a grander scale. To save rivers and fish, he needed people with power, money, and status—anything that would increase his own chances of winning the battle. The ease with which he made this jump-shift was astonishing, but he had always been comfortable hunting for the biggest game. It turned out that people at the top of the food chain weren't unlike fish; they were wary and hard to catch. Like a big trout or salmon, people of power and wealth were not easily fooled, and would rise to the surface only when they thought there was something worth looking at.

In this rarefied hunting ground, Guido's maladaptations served him well. He was like Darwin's nut-cracking finch, able to penetrate resources that others couldn't. His social quirkiness and transparency only served to authenticate his character. He blew into the glass-and-concrete offices of power like a clean northern wind, utterly natural, happy, and free. There was no question that he belonged to another world, one that was distant and, for some, stirred a peculiar longing. For his part, Guido was indifferent to money or power, unless these forces could be parlayed into saving wild rivers. Money in particular said little about a person. Some of the least effective people he knew were people of wealth. He had been exposed to them throughout his life, and had noticed that they spent much of their time entangled with the tinsel trappings, where they lost the big, beautiful picture of life on earth.

Guido's job was becoming increasingly clear. He was there to speak for an ecosystem: to communicate the essential nature of that ecosystem and describe exactly what was threatening it. If he could do this effectively, he found, he didn't always need to ask for support toward its protection; the support materialized. The right people seemed to understand, and Guido was getting better and better at finding them. He found his true niche with men who were happy to leave their worlds behind for a week or two, men who had discovered fly-fishing as an antidote for the disease of modern life with its sedation, anxiety, and clutter. Guido came to them like a revelation, as a pure hunter, with his ancient wiring fully intact. He took them to rivers for days or weeks at a time and offered them a portal to another realm. On the river, worldly accomplishments were leveled and an older rhythm set in, one that was distantly familiar and somehow reassur-

ing. Friendships that formed here went deep, and lives that had become complicated and onerous were made simple once again. On the
river, one was a fisherman, nothing more and nothing less. Men of
importance relaxed, and released the weight of work and expecta-
tion. Here, they were just anglers happy enough to take instruction
from a man who, in this realm, stood above them.

As a teacher and a guide, Guido rarely failed to delight. He was
easy and generous and eager to do whatever he could to help others
see the river as he saw it. He paid close attention all the while, studying people with the same rigor and intensity that he studied a patch of
water, or a salmon holding within it. He watched what they ate and
how often. He watched for temperament and strength; he watched
for passion—what they fed on and rose to the surface for. He saw
how some people fought their way upstream, against the current,
while some held quiet in the deep water, while still others gave into
drift and let the current take them to the open sea. Sitting around the
fire after dinner, Guido asked questions and listened. With every fishing season, his understanding of industry, politics, and business grew.

He knew now that he needed every single person who could help
him. He found that if he was explicit about his vision, and articulated
exactly what he wanted, there was no one he couldn't approach, and
nothing he couldn't ask.

His mentors Spencer Beebe and Peter Seligmann helped Guido establish his footing in this stratosphere of influence, where they were
practiced hands. Like older brothers, they had watched Guido floun-
der during his years in DC, when he couldn't raise a dime. They had
also seen his resilience, and his grit. Spencer saw something else too.
Guido's natural proclivities were aligning with the world of conservation. Such alignments were rare, and they were powerful. As Guido
gained real traction with his stronghold mission, Spencer saw fit to
initiate his younger cousin into an elite circle of fly-fishing friends,
older men who could help him in any number of ways. Spencer's pal,
the writer Thomas McGuane, fished with Tom Brokaw and Patagonia founder Yvon Chouinard. The three of them had formed a trio
they called the "dough boys" because they had the money to explore

the earth's farthest regions with their fishing rods. These celebrity fly fishermen had the power to effect serious change in the world of conservation.

McGuane instantly took a liking to Guido. Spencer had described his cousin as the "real thing," and McGuane agreed. Guido was a character, and his stronghold mission was compelling. When he heard about the family cabins on the Deschutes, McGuane proposed that the dough boys plan a float trip down the river and talk further.

In 2001, Guido took McGuane and Brokaw on a four-day float trip down the Deschutes. The weather was scorching, which made for a warm river and dismal fishing. The conversation, however, was excellent. For someone rarely dazzled by celebrities, Guido found himself starstruck by Brokaw, who had inexhaustible energy and transmitted the same gravitas he had on television. Brokaw regaled them in his deep baritone with stories of his recent trip to Baghdad, referring to world leaders by their first names. When he wasn't covering the news for NBC, Brokaw was on his ranch in Montana, where he fished and hunted year-round. His experiences on the rivers and hills of his huge tract of land had converted him to conservation; he described how he had seen and felt things in nature that had changed him for the better. Over lunch on the riverbank one day, Brokaw invited Guido to pitch his stronghold mission. Guido had been waiting for such an opening, and had taken the care to laminate his map of the Pacific Rim. In the streamside shade of alders, they sat in the tall grass with sandwiches, beer, and the laminated map while Guido painted the picture of strongholds. Brokaw thought Guido's approach was smart and bold. Guido was doing good work, he said, and he'd be happy to support him. Brokaw, a small-town kid, knew the power of names and connections, especially his own. At the trip's end, Brokaw urged Guido to look him up in New York, and they started planning a trip to Kamchatka.

Chouinard made a separate trip to the Deschutes and offered a different model of strength. Quiet and tough, he was not one for chitchat. He had little interest in worldly affairs, especially if they involved politics and big corporations. Chouinard believed the world had to change, but that change had to start from the ground up, with grass-roots activism. Proceeds from Patagonia supported such change, and the brand had done much to bring environmental awareness to the mainstream. But Chouinard himself avoided the spotlight. He had created a highly successful corporation, but this had done nothing to change his way of life, or his values. He lived simply and was still happiest surfing, summiting peaks, or fly-fishing. Like Guido, the most important conversation he wanted to have was with nature. Guido recognized Chouinard as a fly fisherman's fly fisherman; he was hardcore and determined. Chouinard recognized Guido as part of his tribe. In his quiet way, Chouinard became a long-term supporter of the Wild Salmon Center.

Later that year, Brokaw began introducing Guido to people who could help him, vouching for his credibility with the New York intelligentsia. Later he would do more, delivering keynote speeches for the Wild Salmon Center, and appearing as a featured guest at Guido's NYC events, where Brokaw could lend his heft. The two men became friends, solidifying their friendship with two weeks in Kamchatka, and then later on Canada's Dean River, where they fished for a week with Spencer Beebe, Yvon Chouinard, and Robert Rubin, the secretary of the treasury.

It seemed to me that Guido was effectively assimilating the instruction of highly accomplished people. He was demonstrating more refined social skills, evolving into someone who rarely made a misstep in his interactions with others. I remember noting that he was becoming authentically charming—able to laugh at himself and be gracious toward others. He was displaying both humility and humor. At the same time, he seemed more confident and secure operating from his natural high altitude, where he could see the bigger picture. Somehow he was managing to be in two places at once.

Later that year, he met someone who shattered his ceiling for extraordinary personhood. It was an old family friend, Whitney MacMillan, who made the introduction to Supreme Court Justice Sandra Day

O'Connor. After hearing tales of the swashbuckling Rahr in Russia, Justice O'Connor insisted on meeting Guido. She was an avid fly fisherwoman and was keenly interested in fishing in Kamchatka. But first she wanted to know more about the place. Guido was happy to stop by on his next trip to DC a few weeks later. Here, in her chambers at the Supreme Court Building, they swapped stories about clear, cold rivers and rainbow trout. The two warmed to each other instantly. Justice O'Connor saw in Guido what many saw: an incorruptible and fanatical angler for whom the natural world was the most thrilling place on earth. In the middle of what could be a joyless, bureaucratic city, Guido conjured another world for her, one she yearned for. When he left, Justice O'Connor told him he was welcome to visit anytime, and over the next years, Guido stopped in on her when he had work in DC. When he did, she closed her office door so they could talk fish in peace.

Years later, when Guido found out that Justice O'Connor was speaking at the World Affairs Council in Portland, he called her up and offered to take her steelheading on the Deschutes. It was prime steelhead season. Justice O'Connor, who had never fished for steelhead before, enthusiastically accepted.

On an overcast September day, Justice O'Connor arrived at the Deschutes ready to fish. Guido had made all the preparations. He'd asked a fishing-guide friend to bring a drift boat down from Warm Springs, in case the fish weren't biting around camp. Justice O'Connor was eager to get in the water, so Guido got her outfitted. After demonstrating a few steelhead casts on the lawn, he encouraged her to make some casts of her own, explaining how steelheading demanded a different approach than fly-fishing. Then he took her down to the parapet, below Grandfather's cabin. It was still a bit early in the day for steelhead, he told her, but it was a big run that year; who knew when or if the fish would bite.

Guido led her down the bank and they waded into the water together. He saw Justice O'Connor react to the power of the Deschutes, and the tug of the current over the slippery rocks. As they waded in farther he grabbed hold of her wading belt and noted her relief.

"Don't let go!" she told him over her shoulder. Guido was not going to let go. America's first female Supreme Court justice was seventy-five years old. If she fell, she could break a hip—or worse, disappear into the swirling river. Guido held on tight as they started fishing down the run, instructing her to cast across the river and let the fly swing in an arc. With steelheading, you covered the water methodically, sweeping your fly as you stepped downriver, he told her, one foot at a time. "If you get a pull, let me know," Guido said, "but don't pull back too hard on the rod. You could break the line." Guido had put a twelve-pound test on the leader, which was pretty strong. But steelhead were also strong.

They fished down the run, making small talk as Justice O'Connor found her footing in the river. She was getting the hang of it; casting her fly, letting it swing, taking a step, making another cast, letting it swing. It was a warm, cloudy afternoon, and while Guido was enjoying himself thoroughly, he prayed that the justice would see some action. Then she said, "Guido, I felt something."

"Strip in your line and wait a minute. Let's rest the fish. If a steelhead bumped your fly, he's interested. Don't rush it; he isn't going anywhere."

They waited a minute and she cast again, swinging her fly in a perfect arc. The steelhead hit her fly so hard it almost pulled the rod out of her hands. Forgetting Guido's instruction, Justice O'Connor pulled the rod straight up as hard as she could. Guido could see her fingertips pale with tension. She was holding the rod so fiercely, he was worried the line would break. For the steelhead, this was like hitting a wall. It couldn't swim another millimeter downstream against the kind of resistance Justice O'Connor was offering. Nor could it swim down to take shelter at the river's bottom. The only choice the fish had was to burst into the air and cartwheel across the river in explosions of spray.

Justice O'Connor screamed with delight and Guido thanked the river gods both for protecting her line and offering such a beautiful fish. It took them twenty minutes, and a thrilling fight, to land the steelhead. When they got back to camp, the justice asked Guido to pour them both a scotch on the rocks in celebration. Guido built a fire and made dinner while listening to stories about the justice's life and career. She was candid, warm, and funny, and Guido enjoyed every moment with her.

The next day they had a few more bumps but no bites. In the river silence of the canyon, they talked. That evening Guido got up the nerve to ask her if she'd watch his slide show, knowing that such a request might stretch the limits of her good nature. When she agreed, he hung a bedsheet over the window and introduced the justice to his life's mission. Justice O'Connor sat at full attention as she focused on Guido's presentation. Later he thought he had never had such a rigorous audience. Every slide was questioned and every statement cross-examined. The twenty-minute talk stretched to over an hour. After a long day fishing and a dinner with wine, Guido was spent. But Justice O'Connor was ready to deliver her assessment. "Well, Guido, I think you're doing the right thing—and I think it's the right strategy." Then she got up and, despite Guido's protests, insisted on doing the dishes.

In the morning Justice O'Connor woke up and swept out her cabin. Later Guido discovered she had taken pains to sleep only on the top sheet of her bed so that she would dirty only one sheet. The next morning, they floated down the river in the drift boat and fished new terrain. Near the end of the day, Guido ventured one last question. Would the justice ever consider serving on the advisory board of the Wild Salmon Center? Without hesitation, she replied, "I'd love to."

In the following years, Justice O'Connor answered Guido's phone calls and offered him counsel. And in 2010 she would become a critical ally in the epic fight against the Pebble Mine, which threatened to destroy the second-largest stronghold in the Pacific, Alaska's Bristol Bay.

CHAPTER 19

THE HUMAN ECOSYSTEM

G UIDO'S REMARKABLE GAINS HAD a cost. Lee's realization that she would be spending much of her life alone dawned slowly but surely. As the protector of their precious family, Lee was the reason Guido could roam so far. In 2002, they welcomed their second son, Sumner, as well as a new dog. Small children and a puppy meant sleepless nights, constant vigilance, laundry, cooking, and cleaning. This with a husband who would disappear for weeks at a time to places where he couldn't be contacted. The trips to the Russian Far East were the worst; Lee came to hate them. There were so many things that could go wrong. She wouldn't even know if Guido had drowned in some wild river, or crashed in the middle of nowhere in a helicopter. She tried not to think about it as she raised small children alone, overwhelmed, and often broke.

She thanklessly dealt with the many domestic crises that arose on her own. There were rat infestations, remodels, influenza. The stress and isolation wore her down. Some years it got to be too much and she scrapped her stringent financial plan that allowed them to survive on Guido's modest salary and charged airplane tickets on her Visa card, flying to one of her sister's homes in Florida or Montana, or home to Minnesota, hauling baby gear, two boys, and a crated dog. To say that Lee's family was disappointed with Guido was an understatement. They were furious with him. He had taken their beloved daughter and abandoned her. As a father, he was taking terrible risks. To them, Guido seemed self-centered and irresponsible.

Most of the time, Lee held it together. She knew Guido's work was important and she reasoned that her contribution to the earth would be in supporting him. Her own career in conservation, of course, had taken a back seat. The kids came first, the Wild Salmon Center came second, the marriage came third, and Lee's career came last. Lee herself was lost in the shuffle. But she soldiered on, growing tougher with every year. She forged friendships with other mothers, who had more normal lives, and pretended her life was normal too. That Guido was generally doing exactly what he wanted to do, and was no doubt perfectly happy, came as an irritation. Somehow, this was proving to be his birthright. It would have been intolerable had he not been making progress. When she could remove herself from the battleground of her life, she could appreciate that all of his standing in rivers was amounting to something. She missed him when he was gone, or maybe she just missed the idea of him. For when he came home, it was rarely as she imagined it would be. He had the glow of a hero returning from a glorious war. Lee could see domesticity was a shock to his system, and this annoyed her even more.

Each time Guido returned home, not only did he have to be reacclimated to diapers and baby food, but he no longer knew the rhythm of the household. Lee saw his discomfort and his impatience, and she wanted him to go back to where he came from. He was offering her no relief at all. But Guido would adapt, as he always did, and they would settle back into a familiar and contented routine.

Fall meant Guido was in Russia, but spring brought another kind of absence. This was when the spring chinook started coming in from the Pacific, hitting the coastal rivers of Oregon and Washington State. These "springers" were the most powerful and difficult salmon to catch, and Guido was obsessed with them. Lee called it buck fever.

Every year when the snow began to melt, she watched as her husband went temporarily insane. In April, weeks before the fish came in, Guido would start to get twitchy. He'd wake up at three A.M. and fish all day, hoping to catch the first chinook in from the Pacific—and then try to get home in time to help with the children.

There was a reason springers were hard to catch. These chinook swam in schools, deep in the current, rarely surfacing to eat as they made their way inland. Few if any fly fishermen had figured out how to entice a spring chinook to take a fly, but Guido was hell-bent on succeeding. He knew the tactics of the heavy-tackle fishermen who sank their weighted lines and bright lures deep into the current to startle or aggravate the deep-swimming chinook into snapping at their lures. He faced these fishermen in his little rowboat every spring on the Nestucca River. Amused by Guido's fly rod, they tipped their hats and had a little laugh. It was a fool's mission. After a few days, their derision was tinged with respect at Guido's doggedness. He fished for five days straight, living in his Vanagon next to the river, eating elk sausage and bread and beans from a can. He fished sixteen hours a day without a single bite and thought nothing of it.

The gear fishermen on the Nestucca were surprised to see Guido the next year, and the year after. These years ended like the first, and Guido went home to his family unwashed, underslept, and without a single fish. The time didn't matter to Guido. He knew he was getting closer; he was beginning to see the full picture. It wasn't just a fish in a river. It was a fish at a specific point in its life, traveling through a river with hundreds of microenvironments along the effluvial flow of the bedrock. If and when he finally caught a spring chinook on a fly, it would mean he had shaken the hand of an entire ecosystem, and this was what he lived for. His whole being was honed to the task. He thought about chinook constantly, and waited restlessly until spring, when he could fish for them again.

Lee realized her husband was out of control. She had always known Guido was half-wild, but now she wondered if there was something more seriously wrong with him. His connection to nature was charming. But somewhere along the way passion had turned to compulsion. It was the pressure at work, Lee thought. Beyond that, she tried not to think about what it meant, or how she was going to cope with such an affliction in the long term. She focused on the fact that being in a river was good for him, that he loved it.

With the birth of their third son, Henrik, Lee's domestic work intensified, and Guido's absences became harder to live with. Lee would talk to him and could see he wasn't hearing her. He was far away. She tried to fight it, to bring him back to her, to the family. She tried with her innate sweetness and understanding; she tried with diplomacy. When these strategies failed, she tried a dispassionate presentation of the facts. She used every weapon she could find. When none of them had any effect, she blew like a volcano, breaking a plate or two and ranting at the injustice of it all. It wasn't fair! When he wasn't physically gone, he was mentally gone. Did he understand how hard it was for her to have small children, no career, and no money? He did. Guido, who never lost his temper, was distressed by her rage and unhappiness. He understood that he was a challenging partner. But some things he was helpless to control.

Ultimately, Lee surrendered to buck fever, and to all the other fevers that hit throughout the year when Guido bugged out of town and retreated to the coast, or the Columbia. She consoled herself; Guido was saving part of the planet. That he came back from the river happier made it easier for her to let him go. She could see it healed him on some level. In later years, as soon as he got the twitches, she didn't just let him go, she pushed him out. Having him around was too maddening.

Guido chafed a little at their life in the city, which was intensely choreographed with school, music lessons, and after-school sports for the kids; dinners, parties, and events with Lee—along with the occasional date night. While he was immensely proud and adoring of his family, this life did not have the pull of the wild for Guido. On the Deschutes, things were different. Guido was in his element, and happy. His role there was clear. He immersed his boys in the world he knew best. There were dangers on the Deschutes, he told them, but the challenge was to be comfortable instead of being afraid. The key lay in understanding, and in seeing. For instance, the little black spot in the sky above them was an eagle that could see every eyelash on their eyes from two thousand feet away. From where it circled high above, it could spot a snake or lizard lying right next to them.

Guido was fascinated by his sons and their disparate natures. The three boys were all strikingly handsome, with blond hair and blue eyes, but the similarities ended there. His firstborn, Gee, took after his father as an artist and a naturalist-hunter. Blessed with a gentle, self-contained nature, Gee needed no one's approval but his own. He worked long hours on his cast and on understanding the river. Like his father, he hunted not to kill but to better know a world he loved. Sumner, the middle child, was sparky, intense, and plagued with his father's restlessness. A gifted athlete and as fast as the wind, Sumner was driven and challenged his father more than the other boys, demanding more of Guido as he demanded more of everything. Henrik, the youngest, was inscrutably happy and easygoing, with an unusually high social intelligence and his father's dyslexia. As third-born, he had grown up out of the spotlight and was in many ways a mystery to his parents. Guido observed these boys as he observed everything, with a slightly dispassionate eye, and a continual assessment of their biological ability to survive. He was aware that he himself had beaten the system, just as he was aware of the multitude of factors that had aligned for him to do so. It was unlikely that his boys would be so lucky. While sanguine about his own ability to survive, he worried about his sons.

Under their father's instruction, the boys became immersed in the natural world and disappeared for long hours to hunt bull snakes, garter snakes, and blue racers, sweet blue-bellied lizards and fierce alligator lizards. But they observed their father's rules. They were never to grab a lizard by the tail, for its tail would break. They were to hold a snake near its head so it couldn't bite them. If they were bitten by a nonvenomous snake or lizard, they were never to yank their hands away, but to carefully separate the teeth from their skin, so the teeth wouldn't break. There was no reason to fear being bitten, except by a rattlesnake. The boys knew to look out for rattlers when it was overcast and the ground was warmer than the air. They learned to pay attention to the weather, to sense when it flipped, as it often did on the Deschutes, bringing the snakes out from their holes to curl on warm rocks. If they saw a rattlesnake, they were not to panic; they were to calmly walk around it and then tell their father.

When we were growing up, our grandfather's practice had been to kill rattlesnakes within a hundred feet of camp, either blowing their heads off with a shotgun or decapitating them with a shovel. Lee argued in favor of the sense in this practice. Guido, however, chose to "rescue" resident rattlers by catching them and taking them in the boat to the other side of the river. This was acceptable until the day they watched one of these rescued snakes turn around and swim all the way back to its side of the river—at which point Lee won the argument. Guido was not happy with the verdict, as he was never happy about killing snakes, whatever venom they carried.

When they were old enough, the boys learned how to wade the river, staying away from rapids, whirlpools, and drop-offs. Guido taught them how to fish in all conditions, from the densest cover, against a steep cliff, in the thick of trees. They learned to tie their own flies and to attach them to the tippet, toiling over the necessary knots, twisting and looping filaments of nylon into unbreakable lines.

The long, uninterrupted days at the Deschutes healed the Rahrs as a family. As Lee emerged from the heavy lift of motherhood, she resumed a career in renewable-energy and water-resource planning. Campaigning to increase Oregon's use of residential solar energy, she went on to organize rural coalitions to strategize water use to accommodate the changing climate. Lee was now balancing a blossoming career with parenting and Guido's unrelenting travel schedule; it was often Lee who needed to rest now as Guido cooked, cleaned, and attended to projects on their never-ending to-do list. He was up at dawn, and had breakfast ready for the boys while Lee slept in. He led his sons on hikes up Eagle Creek and towed them along when he went fishing. At night, he read to them and told them stories about Russia. When Lee was able to catch her breath, she could appreciate that Guido was a loving and good (if unusual) father. But, for all his ability to be present, he still had an off/on switch, and it could flick to off at any time.

I knew that switch. I knew the feeling of ceasing to exist for Guido. As we brought children into the world at about the same time, I had wondered what kind of a parent he would be. I was not entirely surprised to find that he relied completely on his instinct, on the same ancient wiring that told him where the fish were. I remember the first time I saw it. His firstborn, Gee, went with him everywhere. One winter, when Gee was just shy of two years old, the three of us crossed the Deschutes and drove up a few miles to fish near a spot we called Grandma's Hole. With Gee stowed on his back, Guido followed the short, steep trail down to the water's edge and waded a few feet out. I stayed above to watch and read. It was a sweet scene, the wind catching Gee's bright hair as he peered over his father's shoulder from the backpack, his eyes following his father's rod as it moved gracefully, hypnotically, back and forth above the cold, clear water, the sun playing on the water and in the grasses and trees around us.

Twenty minutes later Gee was getting restless. Guido reeled in his line and waded back to the grassy bank and set him down on a patch of flat ground with some big rocks to explore. Gee happily played on the riverbank, and his father checked on him with a backward glance every few casts. Guido had spotted something, a hold or riffle far across the river that held a fish, and his attention was on getting his fly as close to it as possible. This morning, as on so many others, I found myself absorbed in the harmony of his unbroken motion, by how he became part of the river, the breeze and the current curling around him. It was deeply restful, this merging of man and river. I considered how many of these moments Guido had delivered, that were as whole and perfect as blown bubbles. The moment drifted in front of me now as I became lost in thought.

When I looked away from the river to the bank, Gee was gone. I yelled and scrambled down the bank, spotting the bright blue of his jacket in the river. Gee had fallen and was in the water, his blond hair vanished from sight. Guido had already pivoted and, quick as lightning, had scooped the boy from the glacial water. As they climbed up the bank, neither father nor son made a sound. Guido was grasping his son firmly in his arms, and seemed utterly relaxed. When I made my way to them, I could see that Gee was wide-eyed, drenched, and shocked. He stared up at his father and his father gazed back, mirroring neither panic nor alarm but halcyon calm. The water was cold, but Gee was fine.

We got him to the Vanagon, undressed him, and bundled him in our dry clothes. I kept waiting for Gee to snap out of his trance and cry, or react with some kind of hysteria. He had been submerged in frigid water, weighted down by his sodden winter clothes, unable to move. Surely the shock would wear off and he would fall apart, as children do. But Gee stayed quiet, his huge blue eyes locked on his father. His lips were blue and his teeth were chattering, but with every action and word, his father moved him away from the moment of terror. For Guido, it was already a memory, if it existed at all. I could almost see the moment the father's reality became the son's.

I've thought about this event many times since, about how our children mirror how we are in the world, how we have such terrible power in their early lives. Our boys were lucky on the river, as we had been lucky. There were gashes and broken bones, but this was as it should be. For a good many years, they ran naked and wild with snakes draped around their necks as they spent their days hunting for lizards, frogs, and any small creature they could get their hands on. Guido educated them on every catch, teaching them how to handle captive creatures with respect. My memory of those days is poignant, for our troubles, though new and at times startling, were relatively small. We were standing in the middle of the current, holding strong. We had plenty of questions and grievances and no real idea of what was coming. That we were together was alone reason for happiness.

Maybe even those closest to us don't notice when we change, take new shape, or suffer some gain or loss of personhood. Maybe these things are unknowable by anyone but ourselves. It is with disappointment and sometimes relief that we learn that no one else is watching us that closely, even our loved ones. That even in the midst of the most cloying togetherness, we are alone.

One night after dinner when the boys were asleep, Guido led Lee to the fire and took out a map of the Russian Far East. He spread it on the coffee table, indicating three small yellow circles. They marked the rivers he had helped protect: the Kol, the Vengeri, and the Hoh. In a quiet, proud moment, he told his wife that if he died, he could at least die knowing he had protected these rivers. It was a good feeling, he said. Lee looked at the map and then at her husband as her eyes filled with tears. "I have three circles too," she said. "They're our sons."

CHAPTER 20

THE HUMAN FACTOR

IN THE LATE NINETIES, Guido had brought Spencer Beebe to Kamchatka, and the cousins floated the Oblikovna River. As they drifted down the wild river, they stopped to fish from its banks and talk. Spencer's restless intelligence had driven him on from Conservation International to consider the deeper challenges of environmental protection. It was impossible to protect everything, he explained, but he thought there were ways to harness economic forces to change the system. The fact was, the economy could not be at war with the environment; the two could not work against each other.

Guido listened closely, for his cousin was speaking to the most complex and critical factor in the conservation equation—people. When it came to salmon rivers, people were inevitable; the two species had been intertwined since humans first populated the Northern Hemisphere. People had the power to make or break the salmon habitat, and Guido's primary strategy was to protect the rivers from humans. But Spencer was making an important distinction. It was one thing to protect a habitat from people, and quite another to support people in the protection of a habitat.

Spencer adopted the term "conservation economy." In such an economy, protecting the environment became profitable for its people. A healthy ecosystem provided gifts, and these gifts could make its protection economically sustainable. In the silence and repose of Russia's deep wilderness, Spencer shared his vision for Ecotrust, the organization he had founded to explore this new paradigm, along with models of development that fostered more resilient communities, economies, and ecosystems around the world.

Spencer purchased an old warehouse in Portland's industrial neighborhood and made plans to renovate a three-story, light-filled brick-and-wood building that would house Ecotrust. Soon thereafter, his old friend Yvon Chouinard's Patagonia moved into the first floor. And on the third floor, the Wild Salmon Center found its first permanent offices.

To illustrate the unique dynamics of conservation economies, Ecotrust invested in an emerging mapping technology called Geographic Information Systems, or GIS. These were high-tech maps in which data about human and natural populations and their habitats were fed into a computer and the computer generated an image of the data. The data could be organized any which way; it could create statistics, analyze and search for patterns. It could represent the economic relationship between a place and its people. Unlike traditional maps, there was no limit to the amount of information that could be fed into a GIS map. Guido, who had studied GIS mapping at Yale, saw how powerful the advanced technology would be for Xan and her atlas. Spencer agreed, and Xan was trained on the new GIS system. Soon she was inputting her trove of information into the GIS computer, layering it in an almost geological fashion. What was emerging was a more comprehensive picture of the distributions of all Pacific salmon species—the rivers they spawned in, the places they migrated to, the distances they traveled. One could now see that within any population of species, there were countless subpopulations; the map had made clear the intrinsic diversity that had been key to the survival of the genus.

In her travels across the Pacific Rim, Xan was also confronting the reality of the human factor. She had found that mirroring the salmon ecosystem was a human ecosystem, and the two were inextricably entangled. In fact, she argued, when assessing the health of a salmon population, one also had to consider their human counterparts. How did they treat their fish and rivers? Guido had encouraged her to widen her study, to talk to various people in the northern Pacific Rim's five major countries—the United States, Canada, Russia, Japan, and China—to find everyone she could with knowledge of salmon: fishermen, fisheries managers, scientists, biologists, who all offered their own data points. Every country, province, and community had its own practices when it came to hatcheries and fishing quotas; each had its own standard of river health. Xan collected them all and fed them into the GIS map. Only when the humans had been observed and their practices evaluated could a prediction be made about how well the salmon might survive in the long run.

She also rethought the core geography of her atlas. Because salmon themselves paid no attention to national boundaries, she broke the Pacific Rim into numerous "ecoregions" that more accurately reflected their reality. Gone were geopolitical divisions. The atlas delineations were now the rivers and oceans they swam through, as well as the mountains, canyons, and fjords that held them in the six million square kilometers of land they inhabited.

With the human element removed, the salmon revealed a story deeper than nations and politics; the Pacific Rim was one big, fluid system. As Spencer and Guido watched Xan's atlas take shape, they started referring to the region occupied by the five separate nations as one Salmon Nation.

It was Peter Seligmann who tipped Guido off that Gordon Moore had decided to create a new foundation. It would have an endowment of more than $5 billion, and a wing dedicated solely to conservation. Converted by the American biologist and theorist E. O. Wilson, who advised that "we should preserve every scrap of biodiversity as priceless while we learn to use it and come to understand what it means to humanity," Moore now firmly believed in the value of biodiversity and had decided to invest seriously in its preservation. He was considering various "hot spots" in the Amazon and the Andes, and he was considering salmon. The Wild Salmon Center had taught him a lot about the importance of protecting salmon, but such an initiative would be extremely complex. Moore understood the human factor, which was daunting, especially when one considered the myriad communities across the Pacific Rim. It wasn't until Guido brought Xan and her atlas to Moore's offices in Palo Alto that Moore started to think it was feasible. Xan was attempting to represent the nebulous X factor of people, and her atlas was uncommonly comprehensive. In Xan's work, Moore recognized the data and science he deemed necessary to succeed with a salmon initiative, and he saw how the soft cause of conservation could be structured with hard strategies—with projected outcomes and objectives all linked to the data points in Xan's atlas.

Moore contracted Xan and the Wild Salmon Center to continue their work, and then hired a team to investigate the other groups active in salmon conservation. He wanted to assess all the players as they searched for a lead organization. They would look for the right fit, a group with a strategy, goals, and a scale that most aligned with the foundation's own. There was no guarantee that the Wild Salmon Center's stronghold strategy would be the one they chose.

Guido did everything in his power to make sure it was. Global warming had only sharpened his argument. As temperatures on the planet rose, the best assurance for the survival of salmon was a network of cold, clean rivers. The worst-case scenario was that these oases would be the last places salmon could be found. The best case was that some of the fish would adapt, as they had so many times before. There were signs that it was possible. In the past decade alone, Guido had seen two salmon races in the Tillamook Basin shift their behaviors in response to the warming water. Some populations of spring chinook and summer steelhead were now swimming all the way to the headwaters, where the water was the coldest. Such rapid adaptability gave salmon a distinct survival advantage—strongholds would increase that advantage, potentially allowing the fish to respond to and even withstand the changing climate.

Guido wrote a memo for the Moore Foundation that he hoped would convince them beyond a shadow of a doubt that salmon were worth saving and that strongholds were not just the best but the only strategy. They had growing scientific evidence. Moore's own contributions had helped set up the laboratories where Jack Stanford and Ksenya Savvaitova were delving into the mysteries of perfect salmon habitats, and working to create the most successful Russian-American partnership in decades.

Still, Moore had his doubts. He was not convinced that his new foundation should remain committed to projects in Russia. The country was widely considered the toughest place in the world in which to conduct conservation. With its corruption and rule-of-law problems, it seemed like all the money in the world would not make a difference. Russia was tough, Guido agreed, but he was convinced he could make a program there work. He strongly advised Moore to see Kamchatka for himself, and to witness firsthand the effects of his contributions thus far. In addition, he would have the fishing experience of a lifetime.

Moore agreed to visit Kamchatka in the fall of 2002. The seventy-three-year-old billionaire proved a gracious and unassuming guest. A keen observer, Moore asked intelligent, pointed questions and rarely if ever talked about himself. He seemed to expect no special treatment and instead looked for ways to be of use. At five A.M. in the Anchorage airport, he was the only one to think of bussing the group's breakfast tables. He endured the long flight, the layovers, and the crude Russian helicopters without a single complaint. Like everyone who made the trek to Kamchatka, he was quietly anxious to see its fabled salmon rivers.

Their first stop was the bio-station at the Kol River, where Stanford and Savvaitova had set up field laboratories with the help of Moore's funding. Here, in white canvas tents, Russian and American scientists collaborated on building a biological baseline for the watershed. Stanford and Savvaitova gave Moore tutorials in their respective approaches to fish science. They showed him the cameras they had positioned in the water to record the fish-run size and how they measured the fish themselves. They demonstrated how they could measure the signal given off by nitrogen isotopes in trees, and how profoundly intertwined the two ecosystems were. It was solid, original science, and Moore was impressed.

From the Kol, they choppered east to the Zhupanova River, flying over the vivid foliage of autumn. Guido pointed out the muddy tracks of Kamchatka's new pipeline, the only scar on the otherwise unblemished landscape. For the next five days the group floated the magnificent Zhupanova and fished in the shadow of snowcapped volcanoes. Moore caught the biggest trout of his life and got uncomfortably close to some enormous bears who were just busy enough catching their own trout to ignore him.

There was no question that Moore loved being on the river, and occasionally after dinner by the fire he shared his concern that these places would disappear and the world would become one big strip mall, that rain forests would become golf courses and high-rises. That places like the Zhupanova would be lost. Before they left, Guido arranged for Moore to meet with Kamchatka's governor and President Putin's representative in the region. Even in Russia, Moore was a celebrity, and the Russians were on their best behavior, insisting on their ongoing commitment to protecting the peninsula's salmon rivers. Then they asked Moore for his autograph. After signing his name, Moore added a scrawled request: "Please save the fish."

When the trip ended, Moore concluded that Guido and the Wild Salmon Center were indeed functioning effectively in Russia. Guido had achieved his objective, but this did not mean that the new foundation would fund the Wild Salmon Center or strongholds. The decision would not be made by Moore in any case. It would be made by a crack team of strategic analysts assigned to the salmon project.

Moore found his team commander in a petite, whip-smart woman who left a rising career at McKinsey and Company to explore philanthropy. Aileen Lee had brains, heart, and a particular gift: she understood both people and ecosystems, and she understood how the two interacted. The Moore salmon initiative overwhelmed her at first. It involved a prodigious ecosystem that encompassed countless smaller ecosystems—both natural and human—that, together, made up one of the most productive regions on earth. She came to understand that salmon were a remarkable, irreplaceable resource and that their survival lay in the hands of the myriad communities that relied upon them for their livelihood. All had different agendas and perspectives on salmon, and all belonged to a unique geography. Aileen saw what Guido, Xan, and Spencer had seen, that there was an intricate human community surrounding the salmon ecosystem. And these groups needed support and mobilizing; it was the people, not the fish, who needed managing.

The conservation NGOs would be the boots on the ground, educating and uniting disparate communities. It would be Aileen's job to manage them. It wasn't the nature of NGOs to work together; they all had their particular niches and often competed for the same resources—and now they were lining up, keenly aware that one of them was about to get very lucky. In Aileen's view, however, it was essential that they were united. They each brought something to the table, and Moore was offering enough resources for all of them. When the time came, they needed to remove their blinders and see that the scope of the issues they had to address was as wide as the horizon, and there was a place for everyone.

Before anything, they had to establish a leader. Aileen talked to the groups one by one, as she quickly became familiar with the salmon conservation community. Initially, they seemed like a bizarre group of people who spent much of their time talking about fly-fishing. As she listened more closely, though, she gleaned that all this talk was more than idle banter; it was how this niche community connected to salmon. Fly-fishing wasn't just a sport; it was a relationship. And in the strange equation of salmon, fly-fishing, and conservation, one person stood out. Guido Rahr was unlike anyone Aileen had worked with. His realm was distant to hers, but he led her into it patiently, offering clear, simple explanations delivered with an emotional intensity that demanded her attention. It was through Guido's eyes that Aileen came to grasp the importance of wild salmon and the natural diversity that assured both their abundance and their survival. She saw what salmon gave to the land, and understood the potential catastrophe of their loss. Later, when Guido took the Moore team to the rivers of the Russian Far East, Aileen saw for herself the diversity and abundance Guido had spoken of, how populations of wild salmon utilized every habitat in the river and filled them with life.

Guido also had a financial record with Moore. Lots of intelligent, talented people made inspired pitches for foundation money, but often the money won went toward overhead costs or tinkering with a brand. It was a rare grantee who used funding as they had promised to use it. Guido had put the Moore Family Foundation money directly into play, using it exactly as he said he would.

To make the final decision, Aileen and her team brought the conservation groups together so that they could each make their presentation about how they would approach the issue of saving the salmon. One after another they made their cases about the whats, hows, and wheres of their conservation plans. Guido made his plea for strongholds one last time, stressing the need to look at the Pacific Rim as one interconnected system, and that the protection of the last, best watersheds from California to Japan also entailed supporting the human stakeholders. The Wild Salmon Center was taking the long view of protecting intact, large-scale places for generations to come, and this inclusive, grassroots approach was the way to do it.

The Moore team decided that the Wild Salmon Center was uniquely positioned to move the salmon initiative forward and, in 2005, adopted the stronghold approach and awarded the organization $100 million. The funds would be allocated over the next ten years, and distributed where the grantees saw fit. The Wild Salmon Center would partner with Ecotrust to lead the charge on establishing and protecting strongholds across the Pacific Rim.

It was a massive win, and Guido's head spun with possibilities. He was relieved that Moore would provide oversight, and that it would come from some of the most capable people Guido had ever met. Later, he spoke with reverence about Moore's "dream team" and how they had come up with flow charts and quantitative analyses and something slightly mystical called "strategic frameworks," matrices that linked the workforce to the job and projected outcomes. The Moore team was coming up with its own map.

They were tough on Guido; they weren't going to give him the money until they knew he could handle the job. Toward this end, they suggested some immediate changes. First off, the Wild Salmon Center would need to get their books in order; they had to be prepared for an influx of capital, and the management of that capital. Guido took the opportunity to conduct an inquiry of his own. He gathered his people for a day of conversation. He wanted to be sure that, as a collective, they knew exactly who they were and what they were doing. Only then would they be ready for what was to come. Wisely, he asked Aileen Lee to sit in.

Aileen did what she rarely did and flew up to Portland for the day. She listened to eight hours of talks as the Wild Salmon Center attempted to define itself. At the end, she offered her two cents' worth. The Wild Salmon Center knew who they were, she agreed; they were smart, passionate conservationists and they had a great mission. But, she pointed out, what made them unusual was their leader. They couldn't underestimate the significance of Guido's role. She wasn't sure that Guido himself understood it. For all his charisma and charm, Guido's power lay in the wiring of his hunter's brain, in the pure instinct he had on the river. Being with Guido on the river brought people closer to their own instincts, and activated their own dormant wiring for the hunt. It was thrilling to stalk fish with him, and it was fun. His presence alone encouraged heightened awareness, and the sense of wholeness that came with it. In a fragmented world, such feelings could have the force of a religious conversion, an awakening. This was Guido's extraordinary gift, and in terms of conservation efforts, it was a game changer.

The beauty of it for Aileen was that this alchemy was Guido's sweet spot: what he was good at and what he loved doing. As far as

Aileen was concerned, the key to the Wild Salmon Center's success was simple: Guido needed to keep putting the right people in the rivers he wanted to protect. He didn't need to sell or pitch anything; he just needed to describe the river's ecology and natural history, to reveal its secrets in his own way and language. In essence, all Guido had to do was be himself.

CHAPTER 21

STATE OF THE SALMON

IN THE DAILY PLANET RESTAURANT, on the first floor of the Eco-trust building, Spencer and Guido ate quesadillas and brainstormed, the map of the Pacific Rim between them on the table. The cousins were alike in many ways; they both thought outside the box. Driven by the big picture, they were eager to create and, if necessary, tear things apart. They were unafraid to act, and they preferred to act quickly.

Now, using the map in front of them, they strategized where and how to move first. They had funding and they had hard evidence of a basic reality—that the Pacific Rim was one big ecosystem, and it was unraveling from the southern parts of its range and working its way north. The same things that were happening on the US side were happening on the Asian side. It was looking like a house that could burn down from either end.

It was Spencer who came up with an idea that blew the lid off anything that had been tried before in conservation. The scientific proof that the salmon across the Pacific Rim were in crisis gave them a new weapon—evidence of shared culpability. The crisis was one shared by five nations, and the argument could be made that they all had to be part of the solution. What was needed was something they'd never had before—dialogue, an opening of channels of communication, and the sharing of any and all relevant information impacting salmon. This meant breaking down walls—between countries, cultures, and agencies. Forces that traditionally worked against each other would be asked to work together.

Both cousins felt the urgency; they had a moment in time that might not last. Doors that had been closed to Russia and Alaska were open, and they weren't likely to stay open forever. They had to win cooperation while they could. Spencer thought this would start with shining a bright light on the facts, delivering the science that none of them could refute. Xan's maps made it easy enough to see. These images alone had the power of a wrecking ball.

Guido and Spencer marshaled their troops to start circulating the information they had, emailing and faxing select images from Xan's atlas to the agencies, fisheries, and scientists from the United States, Canada, Russia, Korea, and Japan. Along with the images came an invitation to participate in a transpacific effort to find a way to protect the salmon of the Pacific Rim.

Xan's images were jarring. For the first time, one could see across the Pacific Rim where salmon lived, traveled, and spawned. One could see where wild species were still abundant, and where they had been lost; one could also see the location of hatcheries, the red dots that speckled the map from end to end. Hatchery production had been rising steadily from California all the way to Japan. Guido and Spencer believed this was the most serious threat facing wild salmon. They knew that an excess of hatchery fish hastened the demise of wild stocks; the challenge was to prove it to the world.

The initiative was called the State of the Salmon. It was a call to arms for a region united by an ecosystem, one held together by salmon. The State of the Salmon was like a bell that rang across the Pacific Rim. The request for a sharing of data was met with the understanding that one country holding information back from the others, for whatever reason, would hurt them all. Whatever abuses or mistakes had been made, this was not a time to point fingers or judge.

What followed was unprecedented cooperation, with nearly every agency, fishery, and scientist giving up their numbers, opening their files, and disclosing their practices. Reports were made on the health of rivers, salmon runs, and hatchery production. Over the next year, the Wild Salmon Center collected, organized, and compiled the data and published the findings in lengthy scientific reports, the results of the combined contributions.

What Guido and Spencer hadn't known was that there were other groups, apart from the State of the Salmon, who were also concerned about hatchery production, and that the issue was being quietly investigated by various scientists and groups across the Pacific Rim. Research had come in, but until now, it was unclear what should be done with it. Hatcheries were good business. Any data challenging them met with strong resistance. The State of the Salmon emboldened people to share their findings, creating a safe place for them to discuss these thorny issues and the reality of the situation. In 2006, as the initiative gained momentum, Guido and Spencer invited five hundred people from across the Pacific Rim to attend the first State of the Salmon conference and address these issues together.

Later that year, in Vancouver, BC, Guido stood at the lectern before a packed auditorium. He was momentarily overwhelmed as translators from Russia, China, Japan, and Korea prepared to interpret his words. Looking out at the room full of people, he spoke from the heart, welcoming everyone on this historic day. He told them they were there to avoid making the mistakes of the past. Atlantic salmon had been lost partially because the states and nations of the Atlantic had acted as islands; they had not understood how they were connected, that salmon were a resource they shared. The State of the Salmon was the Pacific's chance to do things differently. They still had healthy stocks of salmon, but things were changing and they were changing fast.

That afternoon, and for the days that followed, scientists from

Korea, Japan, China, and Taiwan presented findings that were added to the data already gathered from the United States, Canada, and Russia. Fisheries discussed their challenges, ichthyologists and river biologists gave their perspectives on what the salmon runs looked like in their rivers, and marine scientists described what was happening in the ocean. Later, Guido described it as a kind of renaissance. The exchange of ideas and information generated tremendous excitement and energy. Old models and perspectives were reconsidered. The shared information linked them in a new geography, and this required fresh relationships and understandings. While the members of the State of the Salmon hailed from different locales, they were part of the same whole. Once they looked at what had gone wrong, they would try to come up with common standards and practices that would halt and maybe reverse the perilous direction they had inadvertently taken.

The bad news was that, between them, they were a rogues' gallery of salmon mismanagement. Asia and Russia were the worst of the bunch. Reports commissioned by State of the Salmon found that the recorded catch in the Russian Far East was 1.21 times the official numbers. Japan's abuses were with sockeye and coho, while China's were with pink and chum. Russia was exposed when Japan came forward with its trade statistics for sockeye, which were significantly larger than the recorded Russian sockeye catch—an excess of 9.7 thousand tons. China was complicit with its floating mother ships that processed everyone's illegal catch, no questions asked.

Then there was the travesty of poached salmon roe in Russia, which amounted to at least 54,000 tons annually. Studies showed that 95 percent of spawning fish could be destroyed in watersheds where poachers were active. Criminal brigades sold their roe in fish markets in Russia, Japan, and China that distributed forged certificates. Illegal trade went into the plant and came out certified. This poached roe was on every menu from Tokyo to Moscow.

Canada and America weren't in a position to point fingers; they had already lost some of their best salmon rivers and served as cautionary tales. It was evident that, to some degree, everyone was guilty.

CHAPTER 22

An Ocean Shared

As the research deepened, State of the Salmon uncovered some ugly secrets. Independent agencies across the Pacific Rim came forward with their reports on hatchery production. It was the North Pacific Anadromous Fish Commission that published findings about huge amounts of chum salmon coming from Japan. The numbers were so high that the commission questioned whether there had been an error in reporting. When questioned, the Japanese corroborated that they had been releasing high numbers of chum into the Pacific Ocean for decades. In fact, they had steadily increased their hatchery production since they had built their hatcheries in 1965. In twenty years, they had jumped from producing ten million hatchery chum to eighty million. However, the Japanese reported, the results of this increased production had been disappointing. They had noted in the hatchery-chum population that there was evidence of delayed maturation. The fish were spending more time in the ocean. Then, not as many were returning, and those that came back weren't as big as in previous years. Japan had elected to cut back production on their own because their chum were coming back smaller or not coming back at all.

The fact that Japan's hatchery fish weren't returning to the hatchery could have meant many things. The most alarming was that the North Pacific had reached carrying capacity for chum, and the fish had run out of food. Maybe, in search of food, Japanese hatchery fish were spilling into the rest of the Pacific. This frightening prospect bore out when an independent group of scientists took samples of chum populations thousands of miles away, in Norton Sound, Alaska. They were surprised to find a high quantity of Japanese baby chum, in fact it seemed that there was little else. Decades of intense production of hatchery fish by Japan had edged out the wild fish clear across the Pacific.

These same scientists followed up with another puzzling discovery. Sampling fish hundreds of miles south, on the other side of Alaska, they found an excessive number of baby pink hatchery fish streaming out of Prince William Sound. These armies of pink salmon were so voluminous, they were competing with the sockeye from Canada's Fraser River, one of the most important sockeye producers in the world. Here too the ocean was being crowded by hatchery clones. This time, Japan was not the guilty party. These fish were being produced in Alaska's own Prince William Sound. Here, in one of the state's most pristine places, large private hatcheries were in full-steam production. While purposely located far from wild fish populations, large scale hatchery production of pink salmon exposed an unfortunate finding. Pink salmon competed for the same resources as sockeye salmon. An overproduction of pink salmon was impacting the wild sockeye population, one of Alaska's most precious salmon stocks.

The problem lay in thinking that a hatchery existed in isolation, but this is how the system was set up. Historically, hatchery managers responded to what they saw coming back to their individual hatcheries. When numbers dropped, as they did naturally year to year, they dialed up production. They were not sensitive to natural cycles, or to how one overactive hatchery could impact the entire salmon ecosystem. Nor did they understand that flooding the oceans with hatchery fish threatened the survival of wild fish. Many of them didn't even understand that there was a difference between the two.

Guido felt that this point was critical and needed to be understood unequivocally. The Wild Salmon Center scientists undertook the editing of a comprehensive volume of twenty-one research reports on the impact of hatcheries. They all supported the same finding: hatchery fish were a serious threat to wild fish. It was their contention that the disappearance of wild fish would be calamitous to them all. One only had to follow the life of a hatchery fish to understand why.

Instead of returning to rivers to spawn, most hatchery fish returned to the hatchery. The rich food of the Pacific they gobbled up was thus not returned back into the ecosystem; much of it ended up in the stainless-steel bins of a factory. As these overproduced fish edged out the wild fish with their sheer numbers, fewer and fewer wild salmon survived to make it back to their home rivers to spawn. Without wild salmon returning to the rivers of the Pacific Rim, the vast ecosystems that relied on their rich nutrient supply suffered. The impact moved steadily up the food chain, from microbes to insects, grasses, bushes, trees, amphibians, birds, mammals, and people. When wild salmon stopped coming home to their rivers, the watch wasn't being wound. Nutrients were flowing downriver and out to the sea, but they were not being returned.

The information generated by the State of the Salmon connected the dots: the impact of hatcheries reverberated through the entire food chain, from microorganisms to humans. The broad strokes of the shared findings were irrefutable and sobering: the ocean wasn't a bottomless food source. It was finite, like every resource on earth. If hatchery fish consumed most of the ocean's food without returning it to the earth, every living organism that depended on wild salmon was at risk.

In places like Alaska and British Columbia, it was the indigenous people who were most directly impacted by the loss of wild fish. These societies had built their villages along salmon rivers, where they had lived sustainably for thousands of years. Salmon were integral to their culture and heritage; the annual salmon migration was the most significant event of their year, and marked time itself. If the current hatchery pattern continued, these ancient native communities would not only lose their primary food source, they would lose the foundation of their culture.

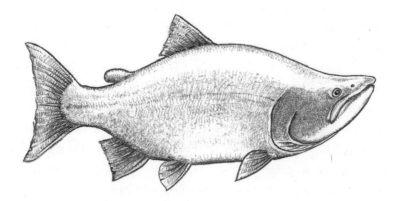

SOCKEYE SALMON

In 2005, the Wild Salmon Center joined Alaska Natives to fight a multinational mining interest that was preparing to excavate an enormous vein of copper at the headwaters of Bristol Bay's most productive salmon rivers. The proposed Pebble Mine would threaten 50 percent of the world's wild sockeye. All the Pebble Limited Partnership needed was a permit. Standing in their way would be a grassroots alliance of Alaska Natives, fishermen, conservationists—and the Wild Salmon Center.

Guido had been chasing salmon in Alaska for decades, but Bristol Bay came to him as a revelation. Tucked in the curve of the southern mainland where the Aleutian chain reached toward Kamchatka, Bristol Bay lay like a gateway to forty thousand square miles of creeks, rivers, ponds, and lakes. It was an untouched wilderness; there were no hatcheries, dams, agriculture, or interference of any kind—and it was home to forty million sockeye salmon.

Years later, Guido took me to see this stronghold in the early fall, when the sockeye came in. If I was going to write about it, I needed to understand. We flew over tundra and rivers in a little float plane, seeing the earth as God might see it, with the original design intact. The trees and forests were a carpet of mauve, yellow, and burnt orange mixed with evergreens. Below were rivers upon rivers, some small and wildly curving, cutting dark across the mossy tundra, others wide and braided and spilling over their edges with the season's high water as they joined in the chaotic tapestry of channels, lakes, and ponds. We pressed against the plane's windows as the sockeye came into view, ruby red in the clear water as they streamed in from the sea. Some spawned in the calm water of rivers, while others traveled all the way upriver to enormous lakes, where they lined the banks like red ribbons.

Landing on a calm patch of river, we explored farther on foot. Immediately I knew this was an ecosystem unlike any I'd seen before. Floating in slow, calm, rain-stippled channels, salmon were everywhere, both spawning and dying. The boundary between life and death here was blurred to the point of disappearing. As riverbeds were filled with new life, there were fish carcasses caught in the reeds and underfoot. In the air was a faint aroma of decomposition as spawning sockeyes porpoised; males chased one another away from redds guarded by females. Guido pointed out a small population of chum salmon that was nearing the end of their spawning cycle and starting to die. Their sluggish bodies were discolored with white patches, half departed already. I realized that this was where the biomass that fed the ocean began, where nutrients from decaying carcasses flowed out from rivers and streams and were carried north by ocean currents. These frigid waters held the ocean's richest soup, where whales, seals, sea lions, and orcas fed. Salmon fed here too and, with their next migration, returned the nutrients to the land to start the cycle anew.

We flew to the Gulf of Alaska to see the fish coming in directly from the ocean, angling so close through a mountain range that the wings of our little plane seemed almost to graze the ice-blue glaciers and lava fields. The red and green gulches and valleys were colored with minerals. We passed over a flock of migrating sandhill cranes, their long, slow-flapping white wings tipped with black.

Once on land, we watched bright silver coho charge toward the coast from the salty waves. We found them holding in pools so clear we could watch them consider our flies. We walked from one river braid to the next, crossing currents so deep and strong we locked arms to steady ourselves. Guido pointed out arctic terns sailing gracefully above our heads with long, arced wings. Farther above, a juvenile bald eagle rode the thermals, testing its balance. We walked down a side channel in the middle of the current, Guido's eyes rarely leaving the water. On the banks, there were signs everywhere of bears: claw prints, broken branches, and the occasional half-eaten fish carcass being picked apart by gulls. Guido paused to assess a pool of coho and in one swift motion stepped into the river, angled his rod, and cast his fly. He hooked a fish immediately, and cursed softly, instantly knowing he had only snagged the salmon's tail. Flicking his rod, he dislodged the fly and turned back to me. It was the briefest of dances, executed without a single wasted motion. I smiled, realizing how happy I was to be back in a river with him.

Rounding a bend, we paused at the sound of crackling branches and grunting. Just upriver, a large brown bear crashed out from the brush into the water, where it sniffed the air and pivoted to face us. For a long moment it stood there, deciding. Then it reared up and smashed its bulk into the river, where it snagged a large coho with its claw and retreated into the dense brush along the bank. Guido watched the performance calmly, remarking that the bears here were so well fed they posed little danger to us. As we continued, the wind stirred the trees, and fish eagles called from above. There were animal tracks pressed into the soft riverbank soil: bears, wolves, foxes, moose. The ecosystem was bursting with life; we could feel its energy, steady and strong as a pulse. At the end of the day, I told Guido I understood why this place was worth fighting for.

Alaska Natives had long known the value of their land, and they knew the value of salmon. As titleholders to 44 million acres of their state's wilderness, they controlled development in Bristol Bay. It was the salmon that convinced them to oppose the Pebble Mine—though it meant passing on billions of dollars of revenue for the state. Salmon were also more than a billion-dollar business for Alaska. More important, the fish kept Alaska's ecosystem healthy and thriving. Having acted as stewards of the land for ten thousand years, Alaska Natives thought beyond short-term profit. Salmon were sacred to them, representing the interconnection between all living things and a wholeness that was intrinsic to their way of life.

The people behind Pebble Mine offered the Alaska Natives millions of dollars and did what they could to curry favor with the locals, promising to improve their schools and healthcare centers—anything to start drilling. Impervious to these material inducements, the Alaska Natives remained skeptical. The Alaska state government, however, proved more amenable. Wide-eyed at the potential mining profits, the state government changed Alaska's land-use designation from 86 percent fish and wildlife to 9 percent with no notice or public proceedings. Most of Alaska's protected wilderness was thus overnight opened up to the Pebble mining project and other "general use." When the Alaska Natives learned of this sleight of hand, they lawyered up, uniting eighteen small tribes to stand firm against the mine and taking the case to court, where much (but not all) of the land-use designation was reverted back to its original designation. Still, the resistance of this small indigenous population was not taken seriously by the mining partnership; it made no sense that a handful of people could hold out against the firepower of one of the planet's wealthiest industries.

The Pebble Limited Partnership crept stealthily forward, managing to stay off the radar of people like Aileen Lee and the Moore team, who had been told not to worry about potential threats to the salmon rivers of Bristol Bay. The spongy tundra was largely uninhabitable, there were no roads, and roughly only seven thousand people lived scattered across the forty thousand square miles. The region was considered impenetrable and impervious to threat. The Pebble Partnership believed otherwise, and continued with its plans to extract the vast reserves of copper from the earth, swearing that this process would not affect the salmon or the surrounding habitat. When the Wild Salmon Center looked into the issue, they found that the claim by the Pebble Partnership was ludicrous. The proposed open-pit mine would be as deep as the Grand Canyon and over two miles long. The mining complex would cover fourteen square miles, and require a new deepwater shipping port a hundred miles to the east, and an industrial road to connect the mine to the harbor, which would cross a national park and countless salmon tributaries.

In 2012, the Wild Salmon Center joined the growing opposition to Pebble Mine and aimed its slingshot, producing an exhaustive scientific report on the effects of the mine on the salmon of Bristol Bay. The results were damning. Eight billion tons of crushed rock would yield only .06 tons of usable copper. The containment of the tailings, or the waste, would require construction of five of the world's highest dams—one at 754 feet. This monumental amount of rock would then have to be covered with water not just for a few years but in perpetuity. This was to prevent the cyanide, hydrochloric acid, mercury, arsenic, selenium, copper sulfate, sodium hydroxide, and dithiophosphate, all used or exposed in the extraction process, from combining with oxygen and becoming toxic. These toxins notwithstanding, even the smallest amount of copper could seep through the mine's earthen barriers, permeating the soil and circulating like a virus, contaminating everything it touched. Copper acts as a neurotoxin on salmon. Even trace amounts of the mineral can disrupt a salmon's ability to navigate, ravage its immune system, and ultimately kill the fish. In the porous tundra of southern Alaska, copper would travel as quickly as water.

Alaska residents joined the tribes and showered the Environmental Protection Agency with letters and phone calls until finally, after 850,000 requests, the EPA conducted its own three-year risk assessment of the Pebble Mine project on Bristol Bay. The findings were definitive. The Bristol Bay watershed possessed "unparalleled ecological value," and damage by the proposed mine would be not only catastrophic to life in the region but irreversible. The EPA used section 404(c) of the Clean Water Act sparingly, but it made the determination now. Development of the mine would stop immediately, and as long as the determination held, it would not restart. The Pebble Limited Partnership responded by suing the EPA, claiming that the government had colluded with environmental activists to create restrictions on mining in Bristol Bay. The judge had no choice but to issue an injunction to freeze finalization of the 404(c), while the Pebble Limited Partnership issued subpoenas to all suspected colluders—including the Wild Salmon Center. The court case was effectively stalled until President Obama, slated to approve the 404(c), left office. When Trump was elected, he promptly settled the case in favor of the Pebble Limited Partnership, allowing the mine to be permitted by 2020. The battle would continue to rage in the coming years, and Alaska would join Russia as one of Guido's greatest preoccupations.

Three years into the initiative, State of the Salmon was achieving its purpose. Gross mismanagement was being daylighted, and imbalances were being righted. Russia alone remained unmoved. It wasn't because they didn't have the data; it wasn't because they were in denial, or didn't understand. Russian scientists and fish managers had participated in exchanges organized by the Wild Salmon Center, and had seen just how terrible mismanagement of salmon could get. They were eager to share the information with their fisheries back home. But when they shared their findings, no one listened. Some were even penalized for infecting their fisheries with tainted Western instruction—no doubt a plot designed to misguide and weaken Russia.

Russia aside, the most pressing task for the State of the Salmon was to create common standards and practices for the management of salmon across the Pacific Rim. It would begin with states and nations agreeing on the amount of fish they would be allowed to catch, how they could catch these fish, and the number of hatchery fish they could produce. The problem was that enforcing these practices required oversight. In the wide-open sea and in the vast northern territories, there was no one to hold fisheries and fishermen accountable. The answer was out there, and it had been created on another ocean, in reaction to a parallel scenario, a catastrophe they all wanted to avoid.

CHAPTER 23

OCEANS AND MARKETS

For centuries, countries bordering on the North Atlantic had enjoyed a seemingly inexhaustible population of cod. These dense white fish filled buckets of fish and chips across the United Kingdom and were popular on menus from New York to Canada, and their livers were processed to make cod liver oil, a nutrient staple in many homes. As populations and markets grew, so did the cod catch. By the second half of the twentieth century, it was clear that aggressive and poorly regulated fishing was affecting the cod population. Scientists issued warnings that present practices could endanger the cod population and the entire marine ecosystem, but the warnings were ignored by government and business alike. In 1992, overfishing and inept fisheries management put the cod population into a steep decline, and one of the world's most productive fishing grounds ran dry of fish. To the utter shock of the Northern Hemisphere, Newfoundland's massive Grand Banks fisheries collapsed. Forty thousand people were put out of work, and the waters were nearly empty of cod. The North Atlantic Ocean had been fished to death. The commercial extinction of the cod fishery was a blow to

Canada's fishing industry, its economy, and the ocean itself. What followed was a trophic cascade that altered the region, both land and sea, possibly forever.

Across the Atlantic, a consortium of businesses and conservationists was distressed enough by this to join together and investigate ways to prevent such travesties in the future. What was needed, they concluded, was an independent entity that could provide oversight; a group of scientists and fisheries experts who could come up with standards for sustainable fishing, and ensure that these practices were observed. But how to enforce such standards? The answer was with a big orange carrot.

The World Wildlife Fund and the Anglo-Dutch transnational company Unilever ponied up the money to found the Marine Stewardship Council. Headquartered in London and with global offices that stretched from Edinburgh to Tokyo and from Oslo to Cape Town, this independent nonprofit set out to create global standards that would ensure the health of the ocean's fish stocks from Australia to Norway. The MSC had lofty goals, but it was grounded in a shrewd, market-driven strategy. Fisheries that complied with the MSC's rigorous standards were awarded a blue label that certified them as sustainable, environmentally sound sources of seafood. In the marketplace, such labels carried value. Increasingly, fish buyers were demanding proof that fish had been caught sustainably. In 1996, the MSC opened itself to the fisheries of the world, offering them participation in a program that would reward them for implementing rigorous sustainable fishing practices. The organization went on to provide a detailed "chain of custody," so that every fish, mollusk, or crustacean in its program could be traced back to its source. Although at first greeted with skepticism by industry and governments, the initiative was welcomed by seafood consumers worldwide, and global populations of lobster, shrimp, prawn, tuna, and krill gained a powerful ally in favor of sustainability: the market itself.

A decade later, the State of the Salmon formally invited the Marine Stewardship Council to help establish a program to promote sustainable salmon management across the Pacific Rim. The Wild Salmon Center would contribute its data and scientists toward the creation of a set of global standards. Of primary importance was that hatchery fish be distinguished from farmed fish, and both from wild fish. Next, the pernicious issues of overfishing and hatchery production had to be addressed. The nations and states participating in the State of the Salmon agreed that it was in everyone's best interest to adhere to whatever standards the Marine Stewardship Council created.

The outlier was still Russia. No one had come close to cracking this nut, even as an independent report commissioned by the Moore Foundation found that "organized illegal fishing in Russia contributed to corruption and criminalization of society, drew fishing income out of the region, and weakened incentives for economic development of the region and sustainable fishing." Cheaters would not prosper—in the long run. But how to legislate to a people who lived day to day, and relied on their black market? In any case, Western markets were closed to Russia. How could the principles of the Marine Stewardship Council find purchase there?

Guido remembered what he had seen with Vladimir Burkanov. The answer to Burkanov's seemingly intractable problem of corruption lay in simply changing the conversation. The Russian fishing industry was strong and lawless, and it was driven by profit. If you wanted to change its practices, you needed to appeal to its pockets. Toward this end, the Marine Stewardship Council had given the Wild Salmon Center a priceless tool. The MSC blue label had real market value. The chain of custody proved that these fish hadn't been poached or stolen or manufactured or tampered with in any way. They were free of the many impurities that had started to affect salmon both physically and philosophically in the marketplace. While capable of producing large quantities of salmon, fish farms were being revealed as dangerous both to the fish and to the consumer. Farmed fish lived in crowded pens that bred disease. To control infection, these salmon were treated with antibiotics. Instead of feeding on the healthy nutrients of the sea, farmed salmon were fed pellets that contained toxins and carcinogenic substances such as polychlorinated biphenyls—all of which went into their meat. For buyers like Peter Pan and Walmart in the United States, blue-label salmon were worth the higher cost.

The higher cost was the key. The Russians had no interest in complying with MSC standards, but what if compliance became profitable? Russians sold most of their illegally caught fish to China, but Western markets paid up to 30 percent more than the Chinese for salmon. What if Western markets opened to Russian fisheries? If Russian fisheries joined the MSC, major American seafood chains might just buy their fish. It took a canny Wild Salmon Center employee a lot of vodka and beer and straight-talking about profit margins to bring the first Russian fishermen around. The promise of more money won the day, and as the word spread, Russian fisheries began signing on to the MSC certification program, a lengthy and arduous test that took at least a year to complete.

But the problem was only half solved. Buyers represented the other side of the equation. Walmart was the world's biggest purveyor of seafood. The American megastore sold a range of salmon, from hatchery to farmed to sustainably farmed to wild. What if Walmart committed to buying only MSC blue-label fish? They were positioned to become environmentalists—and, increasingly, they had a reason to do so. Wild salmon were the most expensive salmon Walmart sold, and the demand for them was rising. But there were fewer and fewer wild salmon available.

By insisting on sustainably caught fish, Walmart could become part of the solution, and benefit in the process. Sam Walton, the reclusive son of Walmart's chairman Rob Walton, had a strong interest in sustainable fisheries. Some years back, Guido had invited Sam to the Deschutes, where they talked at length about the fate of the salmon in the Pacific Rim and the need for sustainable fisheries. Sam was activated by Guido's stronghold mission and had joined the board of the Wild Salmon Center, where he made generous contributions to their sustainable-fisheries program. Guido wondered now if the Walton family would support an effort to recruit Russian salmon fisheries to the Marine Stewardship Council.

A few years earlier, Peter Seligmann had made the case to Rob Walton, who was on the board of directors for Conservation International. Walmart needed to care about the source of its fish. Fisheries were starting to collapse because of abusive overfishing and unchecked hatcheries. MSC compliance could deliver assurances that these abuses were being stamped out, fishery by fishery. Besides, demanding proof of sustainable practices from suppliers was good business. They wouldn't just be getting top-quality fish; they would be preventing the collapse of fisheries all over the world. Walmart could set a new environmental standard in the marketplace, and maybe others would follow suit.

In 2006, Walmart was the first American business to commit to buying only MSC certified fish. It wasn't long before other seafood buyers fell in line. They all wanted what their customers wanted: proof of sustainability from their suppliers. Incredibly, it seemed the *Queen Mary* was turning itself around. The Wild Salmon Center 2009 annual report announced the first MSC compliant fishery in Russia. Every year, the numbers would climb. Fish that were sustainably caught in the cold, clean waters of the Russian Far East could now be bought at Walmart and Whole Foods outlets across America— and salmon populations would begin to recover.

For all its advances, Russia was still troubling Guido. Whatever conservation gains had been won, there remained systemic issues, and poaching was the first. Some rivers were being effectively patrolled by local watershed councils, but poachers who operated in remote regions were well subsidized and had helicopters, armored personnel carriers, and weapons. Fighting them was a sometimes violent and always expensive business, one only deep Russian pockets could take care of. Meanwhile, development of the Russian Far East was steaming ahead; Guido had just returned from the gorgeous Samarga River, where he'd heard about a proposed clear-cutting operation on this remarkable watershed and its extraordinary habitat, one of the most diverse in the world, whose species included Siberian tigers, moose, wolves, roe deer, Himalayan black bears, and brown bears. Not to mention the river's salmon runs. The entire region was

virgin forest. With all that timber, couldn't they cut trees away from
the river?

Guido didn't know how things got done on a federal level in Russia. The Kremlin's system of rule was shrouded in mystery. Much
power was in the hands of the oligarchs, he knew, but how much?
What Guido needed was someone on the inside who could help him
navigate the federal government and possibly intervene against ill-
conceived development plans. He had no idea who that would be, or
how to go about finding such a person.

It required a strange combination of qualities and circumstances to
produce a person who thought nothing of walking up to the military-
industrial fortress that was Russia and simply knocking on the door.

CHAPTER 24

THE OLIGARCHS

P ETER SELIGMANN HAD BEEN COMING to Jackson Hole long
before it became a vacation home for the rich and powerful.
His family had a modest cabin, and every summer they spent the
month of June fishing the Snake River, with the jagged peaks of the
Grand Tetons looming in the distance. Over the decades, opulent
wooden lodges had been erected up and down the river from the
Seligmann cabin, and these belonged to people who could afford to
buy a slice of paradise and build a refuge from the world and its clam-
oring issues. Peace came easily when one was standing at twilight in
a cold, clear river. As did friendship. Bonds formed between the part-
time residents on the Snake, and in 1987 Peter grew close to Harrison
Ford, who had a place five miles south of Jackson Hole. Harrison,
who loved wild rivers, was an impatient fly fisherman; once when
fishing with Pete, he had thrown his tackle into the bushes out of
frustration.

Fifteen miles north of Jackson, and across the river on the south
side, was a big property that had been purchased by an Australian

named Jim Wolfensohn. Peter had heard of Wolfensohn, who would soon become the two-term head of the World Bank. Wolfensohn had been knighted by both his native Australia and by Britain for his contribution to the arts and had served as head of the Kennedy Center and chairman emeritus of Carnegie Hall (where he had performed on the cello). Wolfensohn was also an environmentalist, which made him a potential ally for Peter.

Harrison Ford arranged a dinner for the two men to meet. Both had dealt with big issues on a global scale and had found innovative solutions to complex, entrenched problems. They liked each other instantly. Unlike Harrison, Wolfensohn was an angler, and in the coming years he and Peter developed a friendship while fishing in the jewel-colored water of the Snake River.

Among his many accomplishments during his time at the World Bank, Wolfensohn bailed out a failing Russian economy. Wolfensohn believed in Russia, and he believed in Vladimir Putin, who, when elected president, had vowed to deal with the corruption within the legal, judicial, and financial systems. In his interactions with Putin, Wolfensohn found the Russian president to be an excellent analyst and a man of "great strength and confidence," and Wolfensohn believed Putin would succeed in his mission to get Russia back on its feet. Russia in turn had great affection for Wolfensohn.

When Guido came looking for a way into Moscow's power elite, Peter passed on Wolfensohn's name. Guido wasn't always pitch-perfect, and sometimes when he wanted something he could be too much like a dog with a bone, but Peter trusted his mentee. He thought Wolfensohn would appreciate Guido's passion and be charmed by his transparency. If nothing else, they could talk fish.

Guido immediately started working on getting some face time with Wolfensohn. It took months, but he finally won ten minutes at lunchtime at Wolfensohn's office in Manhattan. Guido took the red-eye from Portland, and one spring morning he stalked the streets of Midtown in his khakis, checkered shirt, and blazer. On the plane he had worked out the briefest possible presentation. Ten minutes would be more than enough time. At noon, he presented himself at Wolfensohn's address, which Guido had located by the limo parked out front.

Wolfensohn was eating lunch at his desk and motioned Guido to come in. His voice was warmed by a soft, singsong Australian accent. The two men shook hands while Wolfensohn assessed Guido with merry brown eyes. Guido immediately felt at ease. Wolfensohn seemed to emanate goodwill, and the undivided attention he gave to Guido was elevating; it was like being in the presence of a benign ruler—a ruler who could talk fly-fishing.

Wolfensohn was eating a salade Niçoise when Guido slid his map of the Pacific Rim across the desk and pointed to the peninsula of Kamchatka and the Russian Far East, an impressive landmass veined with rivers. Then he explained how 40 percent of the world's wild salmon came from these untouched watersheds. Kamchatka's rivers produced one of the earth's biggest sources of wild protein—and they were under siege. Out of ignorance and greed, the Russians were about to repeat America's mistake of destroying ecosystems that could never be brought back. With minimal education and effort, this disaster could be avoided.

Wolfensohn took a long look at the map while Guido continued. Take the plans for aggressive logging on pristine rivers like the Samarga, Guido said. Rivers that flow into the Sea of Japan have the greatest species diversity anywhere, all thriving among old-growth forests. The Russians may not know how essential trees are to salmon habitat. The salmon themselves established these riparian forests with the nutrient deposits of their carcasses. While the trees depended on salmon, the salmon depended on trees for shade and structure— along the banks and in the river itself. The roots, fallen branches, and trees provided key salmon habitat. America doomed its salmon rivers in just this way. Guido ended with a plea. "Our local groups aren't strong enough to go up against industry; we need help in Moscow. You have a chance to do the right thing, Jim. Help me stop the Russians from doing what we did."

Wolfensohn paused and laid down his fork. Then he raised his eyes. "Okay," he said, "this is what we are going to do. I can introduce you to a few people. The first is Vladimir Putin, a close personal friend, but you must think carefully about whether you want to approach Putin. This can help you, but it can also hurt you. The second person is Oleg Deripaska, who is a very powerful oligarch. I strongly urge you to meet with him. The last is Anatoly Chubais, the economist and politician largely responsible for the privatization of Russia."

Guido's time was up. He rose, extended his hand, and thanked Wolfensohn. Then he added, "You should come with me to Kamchatka sometime, and see these rivers for yourself." Wolfensohn said he just might take him up on the offer.

Later Wolfensohn told me that Guido had impressed him as someone who actually might do what he was setting out to do. He wasn't just a dreamer, like so many environmentalists. He was knowledgeable about habitat and passionate about fish, and he understood the politics of conservation. More critically, Guido had a broader view that, for someone like Wolfensohn, was necessary for real problem solving. Wolfensohn backed Guido, and that day marked the beginning of a friendship that would continue into the next decade. Wolfensohn would travel to Kamchatka twice, and come to know firsthand the value of protecting such a place. He would host Guido at his vacation home in Jackson Hole, and offer his young friend both counsel and financial support. Guido was lucky indeed with his mentors, or maybe he just knew how to pick them.

In 2006, Oleg Deripaska had a net worth of $28 billion. He had consolidated his positions and holdings in the consortium Russian Aluminum, during the so-called aluminum wars, in which men vying for control of the lucrative industry "disappeared." According to the US Treasury Department's official designation of Deripaska for trade sanctions in April 2018, Deripaska had been investigated for money laundering and accused of threatening the lives of business rivals, illegally wiretapping a government official, and taking part in extortion and racketeering. There were also allegations, according to the

Treasury Department's Office of Foreign Assets Control, that Deripaska bribed a government official and had links to a Russian organized-crime group. While this was business as usual in Russia, the Treasury Department had repeatedly denied Deripaska a visa to the United States, the free-market heaven where every oligarch romped.

From modest means and trained as a theoretical physicist, Deripaska was at the time Putin's favorite oligarch. He wasn't a fisherman, Wolfensohn conceded, but he had a PR problem and might be interested in improving his reputation in the West. Chubais, Wolfensohn said, would be simpler. Guido left with two phone numbers, the thrill of the hunt coursing through his veins. All he had to do was arrange the meetings, and get himself to Moscow.

A month later Guido was in Moscow eating lunch with C. J. Chivers, a Moscow correspondent for *The New York Times*. The achievements of the Wild Salmon Center were becoming known, and Chivers was interested in their work. It was a friendly, informational lunch in a small sushi restaurant in the center of Moscow, and Guido happily chatted about the dream of an archipelago of protected salmon rivers in the Russian Far East.

In the back of Guido's mind was Deripaska, whom he was meeting the next day. He wished he knew more about the oligarch—or oligarchs in general. How did such men operate, and how much power did they really have? He wasn't sure how much he could ask of Deripaska. He wanted three things at least. One was guidance in understanding the federal government and bypassing the red tape, which was as bad as, if not worse than, it was in America. The second was money for the local conservation groups. Maybe Deripaska would be willing to part with a fraction of his fortune to help protect Russian rivers. Third, and most urgently, was logging on the Samarga River. Did Deripaska have the clout to help with such an issue?

Near the end of the meal Guido asked Chivers about the oligarch. He did so lightly, over coffee. At the mention of Deripaska's name,

Chivers stopped eating. "Oleg Deripaska? Yes, I know of him. Every-one does."

"Really?"

"Guido, do you have any idea who this guy is?"

Guido kept his tone light. "What I know is that he's a big oligarch. I'm hoping he can help us stop the logging on the Samarga River in the Russian Far East. What do you think?"

Chivers shook his head. "Guido, Oleg Deripaska is so powerful that with one phone call he could stop the logging on the Samarga, strip the timber company of their logging rights, and have them all deported."

Guido was both heartened and chilled by Chivers's response. Approaching an oligarch and simply asking him for things was an act that existed outside the realm of sanity. Chivers communicated this clearly. But Deripaska had agreed to meet with Guido for a reason: the oligarch must have needed something too. Whether Guido could provide it for him, he didn't know. Deripaska's sketchy past had opened a door, and Guido was going to walk through it. Guido had something else in his favor: he wasn't Russian. He did not know fear or paranoia; he did not live within an impenetrable system. What was more, he had nothing to lose.

His meeting the next day took place at one of the satellite airports outside Moscow. Guido left the hotel early to make sure he got there in plenty of time; this wasn't a man one kept waiting. When he arrived at the airport, which seemed more or less like a fancy private airstrip, he was informed that Deripaska was flying in from Asia in his Gulfstream V-SP. Guido was made comfortable in the well-appointed boardroom and introduced to Deripaska's relevant staff. Guido counted sixteen people. Some were foundation people, some were public relations people, some were communications people, and some were not identified. The pre-meeting with Guido was over quickly, leaving a silence that was broken every few minutes when someone announced, "Mr. Deripaska will land in twenty-two min-utes. . . . Mr. Deripaska will land in seventeen minutes. . . . Mr. Deri-paska will land in ten minutes."

Deripaska arrived right on time. He was big boned and tall, and his pale blue eyes were large, keen, and on the cool side. The oligarch sat down, crossed his long legs, and said, "Okay, Guida, what do you want from me?"

Guido, ignoring the mispronunciation, presented his map and explained to Deripaska the importance of salmon conservation and how, in the Russian Far East, the most pristine salmon ecosystem in the world, the decision to save the watersheds was a Russian one. Which was why Guido was here. He then told Deripaska that he needed a few things from him: his money, his contacts, and his political clout to navigate the Russian federal government and provide more support for the local conservation groups. He added that he could also use help setting up some sort of association that would promote salmon conservation in Russia, something that appealed to the power elite. If they could be brought to fly-fishing, they could be brought to conservation. Lastly, he needed him to help stop the logging on the Samarga. Then he shut up while Deripaska studied the map. After some long moments Deripaska looked up and asked, "Guida, who chooses the rivers?"

The question was pointed. This was the central issue for Deripaska, and perhaps the only issue. Because Deripaska was Putin's favorite oligarch, he could not be seen to misstep, especially when it came to Americans. Guido was asking for a lot, or maybe he wasn't. Either way, Deripaska seemed willing to take a chance if the favor didn't come back to bite him. Guido stood on the firm ground he had always stood on: Russian fish were Russia's concern, and what bound their two countries together in this effort was science. He held Deripaska's eye and told him confidently, "Our Russian scientific partners choose the rivers."

It was enough for Deripaska. He thought for a moment and said, "Okay. I will help you. This is what I'm going to do. I'm going to introduce you to my guy who can help you with the Russian federal government agency and help set up this new organization. I'm also going to give you some introductions and some funding. But the main thing I am giving you is this guy who works for me, and he's going to help you do all these things." Guido left, cautiously excited that Deripaska's words would amount to something.

The next day, Guido met with Anatoly Chubais in a gray office building tight with security. As the architect of Russia's free market, Chubais had been blamed for the tanking economy and had since survived multiple assassination attempts. Blond, fit, and fluent in English, Chubais listened carefully to Guido's proposal to set up a Russian salmon partnership in Moscow. Unlike Deripaska, Chubais was familiar with Kamchatka's rivers, and was concerned about the pipelines being laid across western Kamchatka, starting at the mouth of the Krutogorova and running south across the peninsula's best salmon rivers. Chubais knew that this pipeline and the roads it required would pave the way for poachers, which would be a travesty for the salmon. Guido's suggestion to involve powerful Russians in the protection of their salmon made perfect sense to Chubais, who committed $50,000 on the spot.

With seed money, an office in Moscow, and inside help, Guido, his colleague Andrei Klimenko, and Serge Karpovich's cousin Gennady Inozemtzev started setting up the Russian Salmon Fund. Genna, who was full of useful ideas, made a logo and flag with the Russian colors. Soon they had a staff of one and were hosting fly-fishing demonstrations and events a few times a year, importing Western experts to show how the sport was done. Available at these events were pamphlets about river lodges and trips to various Russian rivers, with alluring photographs of clear waters and grinning fishermen posing with their quarry. There were also informational packets on how to run a hatchery, which included some commentary about how hatchery fish were not a remedy for the loss of native stocks. Guido's quiet campaign in the heart of Moscow had gone on for a year or two, when it caught the attention of a man Guido had been trying to find for years.

CHAPTER 25

THE INNER SANCTUM

THREE YEARS EARLIER on a trip down the Rio Grande in Argentina, Guido got to chatting to his guide. Hearing about Guido's exploits in Russia, the guide let slip tales of a fly-fishing Russian he had taken down the same river a year or so earlier. Guido's ears were pricked, for this was the first he'd heard of a member of Russian's financial elite with an interest in fly-fishing. Pressing the guide, he learned that this was a dedicated angler who had chartered helicopters to explore the remote rivers of the Russian Far East. The Russian's name was impossible to procure, no matter how much Guido badgered. Beyond a few tantalizing details, Guido discovered nothing. He kept thinking about him, though, and how one such person might change the game for the salmon of the Russian Far East.

As Guido's contacts in Moscow grew, he tried to home in on the identity of the phantom Russian fly fisherman. He cast his line in all directions, trying not to appear too eager; he didn't want to spook a man who obviously valued his privacy. Guido quietly put the word out within the Russian Salmon Fund. Did anyone know of a Russian with a passion for fly-fishing? In the end, the phantom fisherman found Guido, for news of this Russian Salmon Fund had reached the man himself. He was curious about how such an organization had sprung up in Moscow, and intrigued that it was the brainchild of a master fly fisherman from the West. Guido finally had his name.

While Ilya Sherbovich was one of Russia's youngest and most successful entrepreneurs (he would one day control a majority in Russia's equivalent to Facebook), his true passion was fishing. He had journeyed to rivers all over the world, casting his line for freshwater and saltwater fish, and was by all accounts a master fisherman. At the time, Sherbovich was the head of investment banking at Deutsche Bank in Russia and dealt with those closest to Putin. He was intelligent, progressive, and shrewd. He was also decidedly apolitical. This, Guido thought, could be a good thing.

After years of heavy-tackle trophy fishing, Sherbovich had been introduced to fly-fishing in the early nineties by an American colleague while on vacation on Lake Baikal, where they fished for char and grayling. Sherbovich instantly loved the challenge and elegance of fly-fishing, and had since been pursuing trout and salmon across Russia with his light graphite rod. When word of Moscow's little Russian Salmon Fund reached him, he was intrigued enough to accept a meeting with Guido. On his next trip to Moscow, Guido made his way to Sherbovich's offices on Sadovnicheskaya Prospekt.

Sherbovich was a stocky, broad-shouldered man with sharp brown eyes, and he couldn't wait to discuss fishing with someone who knew what they were talking about. While the two men got along like a house on fire when talking fish, the subject of the Russian Salmon Fund was another thing. Guido's aspirations for the organization elicited an eye roll and an undiluted dose of Slavic cynicism from Sherbovich. The notion that fly-fishing and salmon conservation would ever get traction in Russia was nonsense. Guido needed to wake up. For most Russians, fishing meant driving to a river with a few bottles of vodka and passing out on the bank by noon.

Nevertheless, Guido and Sherbovich became friends, and corresponded in the coming months. When Guido came to Moscow they met and talked more about the threats to Pacific salmon. Sherbovich, who identified himself as a conservationist, declared his commitment to helping the situation in Russia. Demonstrating as much, he pledged $100,000 to the Russian Salmon Fund and then diplomatically offered to take it over. Guido's team knew it wasn't so much an offer as a mandate. Russian relations with America were cool and getting cooler, and an American-funded operation in the center of Moscow, once marginally acceptable, was now objectionable. Guido and his team were happy enough to release control of their creation; the Russian Salmon Fund was re-created as the Russian Salmon Partnership, as founded by Ilya Sherbovich, Alexander Abramov, and Vladimir Rybalchenko (who later bought Farlows, England's high-end fishing and hunting store). The Russian Salmon Fund had done what Guido had hoped it would do, embolden powerful Russians to take care of their own rivers and fish.

The more Guido got to know the young magnate, the more he thought Sherbovich would be an ideal member for the board of the Wild Salmon Center. Pragmatic but open-minded, Sherbovich could represent the modern Russian perspective. He could also give Guido a toehold in the icy wall of Moscow politics. It was a stretch to ask a Russian to ally himself with a Western NGO, especially as political relations soured, but Guido resolved to try.

Sherbovich had just purchased the lease for a river on the Kola Peninsula in northwestern Russia. This was the world's best place for Atlantic salmon fishing, and its exclusive lodges and camps were frequented by an elite clientele. Sergei Ivanov, the former KGB spy who was then President Putin's chief of staff, Prime Minister Dmitri Medvedev, former US vice president Dick Cheney, former Federal Reserve chairman Paul Volcker, and President Jimmy Carter had all been guests at the camp on the Ponoi River, where a week's lodging could exceed $13,000. That fall, Ilya invited Guido to be his guest on the Ponoi. He had a plane pick Guido up in Moscow and fly him to Murmansk. It was a big private jet, and Guido was the only passenger. In Murmansk, he was transported in an Mi-2 helicopter to the camp on the Ponoi.

Perhaps the most significant discovery Guido made on the Ponoi had nothing to do with fish. In the company of the Ponoi camp's elite international anglers, he caught wind of another Russian he'd been stalking, one he'd almost given up trying to find. Guido had first heard Alexander Abramov's name many years earlier from Misha Skopets. He'd been trying to meet the man ever since, for here was the Russian who held the keys to the river Guido wanted to fish more than any river on earth. The Tugur River had been on Guido's radar from the moment Misha Skopets had suggested that this river filled with ancient giant fish might be one of the more unusual strongholds he'd ever seen. The Tugur and its enigmatic Siberian taimen had tormented Guido since. He had tried every way he could think of to get there on his own, pursuing any lead that might help him reach this remote watershed. While Guido could get himself most anywhere in the world, the Tugur had remained mercilessly out of reach. He had never come across a river so physically challenging to access. From his calculations, it required a fifteen-hour ride on a primitive road that ended in an equally primitive settlement in lower Siberia, a tiny town called Briakan, consisting of a scattering of wooden houses and a muddy field that occasionally acted as a landing pad for helicopters traveling even farther north into the roadless wilderness. Guido was willing to do all of it, but the Russians had consistently rejected his requests for access.

When he learned that oligarch Alexander Abramov had purchased the lease to the Tugur watershed and now controlled a hundred miles of wild, taimen-filled river, Guido began systematically researching Abramov, looking for any connection that would collapse the distance between them. An initial foray did not give him hope. Abramov was a former plasma physicist turned industrial magnate. A part of the second generation of modern oligarchs, Abramov had been granted few political favors, and had made his own fortune. He and his partner, Roman Abramovich, had amassed the largest steel and iron empire in Russia, employing 125,000 people and controlling about 22 percent of the country's total steel output. Their holdings had an annual turnover of $20 billion. Guido was surprised to find that they also owned Oregon Steel Mills, and had upward of two thousand employees in Guido's home state.

Through the grapevine, Guido learned that Abramov was not only a fisherman but also a trophy hunter. No doubt he had found the ultimate trophy hunting ground in the Tugur. Word had it that Abramov had built a fishing lodge far upriver, reachable only by helicopter. Every fishing season, he welcomed guests, but he kept his circle small, inviting only his close friends, business partners, and other fishermen. It was on the Ponoi that Guido found out that a member of this small circle was Ilya Sherbovich.

It was when Ilya agreed to join the board of the Wild Salmon Center, in 2012, that Guido strategically disclosed his dream to fish the Tugur. Ilya, like an angel from heaven, replied that he'd received an invitation to fish the Tugur by his friend Alexander Abramov that coming September. There was one spot left on the trip—would Guido like it?

Maybe it shouldn't have been a surprise; Guido had penetrated the highly selective fraternity of Russia's fishing elite, and it was an intimate group. But the invitation to the Tugur was particularly fortuitous. Alexander Abramov was conducting an inquiry designed to answer a central question: Could the taimen of the Tugur be caught on a fly? Guido couldn't believe his luck. It seemed Abramov knew that the best sport-fishing rivers in the world were fly-fishing rivers. Was the Tugur possibly one of these rivers? Abramov himself was not a fly fisherman. He found the sport not only uninteresting but inefficient; fly fishermen worked like dogs, and often for paltry returns. But if the Tugur could be established as a fly-fishing river, it could be deemed one of the best in the world, offering access to the highly sought-after "river tigers," Siberian taimen. And this designation interested him very much.

The group Abramov had invited in September 2014 included some of the finest fly fishermen in the world, men who were happy to come from South Africa and Argentina to see what they could wrangle from the Tugur. There were also some novice Russian fishermen, friends from Moscow who had discovered fly-fishing and, like Ilya, fallen in love with the sport. Eight men found their way to the south-eastern city of Khabarovsk. Here, they were picked up by Abramov's Mi-8 helicopter and flown to the Tugur.

The only person missing from this who's who of elite fishermen was Abramov himself, who would be standing by in Moscow. It would be up to the other men to crack the fly-fishing code of the Tugur. Abramov expected regular reports. Perhaps only he knew how hard it was to pull a hundred-pound taimen from the Tugur, even with heavy spinning tackle. It had taken him years to understand where in the river to cast his lure. Hooking a taimen was one thing, landing it another. Even with some of the most talented anglers in the world, Abramov had his doubts about the success this group would have.

CHAPTER 26

THE TAIMEN CODE

ROM KHABAROVSK, IT WAS a long helicopter ride in the most luxurious helicopter Guido had ever flown in. The clean interior and the souped-up engines were a far cry from the gutted-out choppers he had crouched in before. It was the autumn of 2014, and the birch, alder, and cottonwood forests below were bright with yellows and oranges. The terrain glistened with wetlands and lakes. The landscape seemed to be endless, the gleam of luminous trees and water stretching as far as the eye could see.

Abramov's lodge was on a tributary that fed into the Tugur, called the Konin. It was an opaque river, broad and deep and edged with golden birch. Fourteen of Abramov's staff met the group of anglers on the helipad, took their luggage, and welcomed them to the lodge. This was not, as Guido had expected, a rough field camp like they'd had in Kamchatka. There were vegetable gardens and food supplies and domestic animals to be butchered and eaten along with the occasional moose or elk. The staff was highly trained and organized and ready to transport the fishermen with helicopters and premium jet boats. The compound itself was well appointed and masterfully built

out of old-growth pine. The rooms were spacious and tastefully decorated, and there was one for each guest.

The cooking lodge where they would dine featured an oversized
photograph of Vladimir Putin on the wall, posing bare chested with
a fishing rod in his hand. Hanging on other walls were photos of the
biggest trout Guido had ever seen, of taimen caught on the Tugur—
monsters that weighed eighty, ninety, or a hundred pounds. Abramov's
guests gathered around and stared at the pictures in disbelief, electric
with anticipation.

In preparation for the trip, the more experienced fishermen had
devoted themselves to tying flies. The first night they got together and
compared flies and tackle and any experiences they'd had that might
inform them on the Tugur. They had been warned that the river did
not give up many fish. Even with spinning gear, anglers here caught
only one or two taimen a day.

The next morning they broke into groups, with two people per
boat and a guide. They jetted off to various points along the Konin
and the Tugur, places that had yielded fish before. The anglers fished
all morning, broke for lunch, and fished until sunset—and caught
nothing. The next day and the following they fiddled with casting and
retrieving, switched out their flies, and tweaked their approaches, but
they could not persuade a single fish to bite. Sometimes they saw
taimen appear behind their flies, and occasionally an enormous fish
surfaced right next to the boat. The fish were curious enough to follow the flies, but for one reason or another they refused to take them.
The insult to the injury was that they lost their valuable flies right and
left to logs. After dinner on the fourth night, Guido got out his fly-
tying vise and his bags of feathers and fur. Some of the less experienced fishermen had the wrong kinds of flies, and the others would
need their supplies replenished.

The next days were an exercise in total frustration. Even Ilya, possibly the best taimen fly fisherman in Russia, could not catch a decent
fish. Some big fish were hooked, but all were lost. The big mouth of
a taimen was difficult for a fly hook to penetrate. You had to pull
hard to set the hook, but too much of a yank resulted in broken lines,

or an angry fish that had no trouble breaking itself off. As if to torture the anglers, taimen rose briefly to feed on smaller fish, the water swirling and roiling beneath the surface.

The guides alone were not surprised. This had been the pattern on the Tugur. It turned out that this crack collection of anglers wasn't the first to try their luck fly-fishing on Abramov's river. He had flown out other experts from Scotland and England, the birthplace of the sport, and they'd also had difficulty hooking these big fish. More than a few fishermen had been invited to come with their fly rods, and after days of catching nothing, finally gave in to a spinning rod. But Abramov had not given up.

One afternoon Guido's boat motored past what looked like a dead taimen lying on the bank. After passing the fish, Guido requested that they swing back to take a closer look. If it was a taimen, Guido wanted to extract the otolith bone from its head. The Russian guide didn't understand why the American fisherman wanted to look at this dead and stinking fish, but he took him back to what indeed turned out to be a rotting taimen. Guido had the guide sidle up to the bank so he could get out. He took a look at the fish, which was of medium size, pulled out his knife, and cut off its head. Later at camp, he sat outside and split the head with a large hunting knife and a wooden mallet, removing the otolith bone from the center of the skull, a trick he had learned from Misha Skopets. The Russians watched, not knowing what to make of the strange act of barbarism.

Over meals and in the sauna, Guido got to know the other Russians. They were an educated and informed group, and they cared about the wilderness. Guido realized he had found what he'd been looking for during the past decade or so—Russians who understood the value of pristine rivers. The most approachable was Nikita Mishin, who seemed more like a professor than a businessman. Soft-spoken and deep thinking, Mishin had discovered fly-fishing relatively late in life, when he'd gone to Kamchatka with his old friend Sergei Pavlov. Coincidentally, it had been one of Pete Soverel's exploratory trips, just after Guido had come on board. Fly-fishing had been the strangest, most unexpected discovery for Mishin. Even after three days of hacking through logjams and being choppered out to relocate to a rustic cabin belonging to a Koryak herder, he'd found the experience life altering. Later he described it as a revelation. "It was like I escaped from a reality I was living in for a long, long time," he said. First there were the waders, ridiculous rubber contraptions that started with booties and ended in a chest-high bib. Mishin thought his friend Pavlov was teasing him; one didn't actually wear such things. Soon he understood their value, and how they allowed him to enter another world without freezing to death. What impressed him next was how, on a salmon river, there was no boundary between life and death. He watched fish spawn and saw that while at the peak of their lives they were also starting to die. They were transforming before his eyes, moving from one part of nature into another. Mishin realized for the first time in his life that nothing died, that there was only what he called a "fantastic collaboration." He had since seized any opportunity he could find to get back into wild rivers and feel the peace of the continuum that he now understood himself to be a part of.

Mishin recognized Guido to be a special kind of person. He had what Mishin called "soft strengths," which Mishin recognized could be more effective than the traditional Russian machismo. He noticed that while the rest of the anglers were chatting or playing cards after dinner, Guido was tying flies. As their unlucky fishing streak continued, Mishin realized that every day Guido was studying the conditions and the river and adjusting his formula of weights, feather, and fur. He was experimenting with color and flash. While in the river, Guido was rarely at rest. He was observing and analyzing; his mind never stopped working. And at the end of every day he had new ideas for how they might catch these river tigers, and these ideas went straight into his flies, painstaking creations that another might have secreted away for personal use but that Guido was happy to share. In the evenings, Mishin would come to expect a visit from Guido, who would present him with flies he thought might work better the next day.

It was about two-thirds of the way through the week that the seemingly unbreakable code gave way. Mishin thought it was only right that the fishing gods had favored Guido.

The night before, Guido had tied a batch of new flies. He had chosen a tube fly, and it was about eight inches long with weighted eyes and a head made of spun elk hair. It had a long brown wing and a bit of flash in it. On the end was a stinger hook, a single hook attached to a loop of monofilament line that hung toward the end of the wing. Guido had started attaching these hooks, with the reasoning that no matter how tentatively the fish grabbed the fly with its mouth, it would end up with the hook.

It was a sunny day and unseasonably warm. Guido was fishing a side channel obstructed by big logs, doing his best to fish around the logs without losing another fly. Casting across the water, he let his line swing. When it began to straighten out, he started stripping it in, little by little. On about the third pull, something stopped his fly. Guido knew right away it was a big fish. He set the hook, and for once the fish stayed on. They fought in the small channel as the water churned and broke over the body of a taimen that was respectably big. As Guido slowly brought the fish in, his Russian guide was beside himself; this was something he had never seen—and had never thought he'd see. The taimen was forty-three pounds, one of the biggest Russian taimen ever caught on a fly. It was a sign that it was possible, that the code could be cracked.

The trip to the Tugur yielded many treasures. Nikita Mishin agreed to sit on the board of the Wild Salmon Center, joining his friend Ilya Sherbovich. Later that month, 200,000 acres of the Tugur River won regional protection. The only significant failure had been not getting to meet Abramov himself. Guido longed to find out more about the man who had built such a magnificent lodge, who had such excellent friends, and who had come so far in cracking the heavy-tackle code for taimen. He didn't know if he'd have the chance. Too many insults from the West had motivated Putin to shut down diplomatic relations. No one knew how far the Russian president would retract, but for the past two years, all the signs were that Russia was closing up.

CHAPTER 27

THE DISPOSSESSED

THE DAYS ON THE DESCHUTES were short in late November; the sun disappeared behind the rimrock at four o'clock, and the air grew chilly soon after. One by one, our families drifted into the warmth of our cabins and gathered around the fireplace. Books and cards and guitars were brought out, and our boys, big now, sat close with their heads bent, sharing secrets we no longer knew. Guido's son Sumner and my son Elliott were just months apart in age, and as thick as thieves. They were an even match of temperaments and physicality, and entertained each other thoroughly. We delighted in watching them grow up together.

As we edged into middle age, Guido and I grew even closer. Over the years, I had cautiously begun seeking his counsel, and my trust in him had grown. I realized he understood me, and he understood my difficulties. He had been paying attention for some time and had evolved into an excellent student of human nature. When I presented him with a tangle of trouble, he patiently helped me untangle it, offering pithy observations that cut to the heart of the matter, sometimes brutally, but rarely off the mark. I knew by now that I was going to try to tell his story, and our conversations had taken on more purpose.

I was seeing the world differently. On my daily walks, I now hiked high enough to glimpse the Pacific. I often stood and imagined the life that lay below the surface, the salmon that streamed in and out of the bay, swimming tirelessly as we slept. Sometimes I saw it all from the sky, like Guido did. The more I learned from him, the more I understood that beneath the layers of concrete, wood, and glass covering the earth was an intricate network of creeks, streams, and rivers that veined the land in a circulatory system of water as complex as the one that moves the blood through our bodies. Beneath the roads and cities, water was flowing in perpetual motion, moving from land to ocean to sky and back to land. The salmon followed the water, traveling between fresh and salt water like the earth's red blood cells, delivering nutrients from the ocean deep into the interior as they completed their annual migration; nourishing soil, insects, birds, animals, and forests. It was beautifully perfect, and somehow I had become committed to making others see what I saw.

I shared such musings with my cousin, who had led me on this path from the start. Over Thanksgiving weekend in 2013, we sat at the breakfast table and talked. These days Guido had the air of a soldier reporting from the front lines. He told me that things were getting tougher in Russia; it was becoming difficult to sustain operations. Colleagues and key staff of other environmental NGOs were being barred entry or harassed by the FSB. Wild Salmon Center employees were regularly detained and interrogated. One by one, Western NGOs were pulling out of their projects in the Russian Far East.

At the same time, Putin declared the prioritization of the development of the region, steamrolling anything that got in the way. As the Russian president committed more capital to extracting the resources of the Russian Far East, environmental voices were silenced, one after another. Efforts to set aside watersheds in Kamchatka were frozen. Extracting coal, oil, and natural gas now took precedence.

The previous year, the Russian president had delivered his coup de grâce by passing the "foreign agent" law. Under this law, Russians who accepted funding from Western organizations were required to register as foreign agents, or enemies of the state. The Russian colleagues of the Wild Salmon Center were now considered spies. The Wild Salmon Center had been forced to pull their financial support from the groups and individuals that had been working tirelessly to protect salmon in the Russian Far East. These disenfranchised people now had no resources, and they were alone.

Guido looked out at the river, where the sun was catching the water. He told me that a few brave Russian partners had resolved to continue their conservation work on their own—and there were the watershed councils, ragtag groups that, against all odds, seemed to have taken root in communities across the far east. Sakhalin Island in particular was experiencing some kind of awakening. Sakhalin had become a resource colony of the "smash and grab" order. Its people owned little, had tolerated much hardship, and for decades had stood by as multinational companies gouged their hills, clear-cut their forests, and drilled their ocean floors. They had watched silently as nature disappeared and the island's ecosystems were broken, one by one. It was the salmon that pushed them over the edge. Salmon had once been a mainstay of Sakhalin's economy, but the dissolution of the Soviet system had left the factories abandoned and the fish unprotected. Now poachers had taken over, and the salmon were being slaughtered, gutted for their roe, and left to rot in massive piles along the riverbanks.

It was a motley crew of teachers, retired policemen, music teachers, factory workers, and students who came together to fight for their fish. Education, they decided, was the most critical piece. The consensus was that 90 percent of the Russian population knew nothing about salmon. If people understood how much salmon contributed to the earth, things might start to change.

The Wild Salmon Center watched from afar as Sakhalin's councils wrote pamphlets and printed posters. Ads were run in local newspapers. They all said the same thing: poaching and overfishing had to be stopped. Volunteers were sent to visit schools and talk to students about how salmon lived and died, and how much they gave to the ecosystem, nourishing the rivers and forests they all loved. Programs were set up for children to get out on the rivers to see the fish in the various stages of their life cycle. Across the generations, from river to river, people began to understand that salmon swam one way when they were young and another when they were older. Whichever way the fish were swimming, they needed to survive.

Guido lit up when he told me that there was a rumor that the Cossack community had joined the watershed council in northern Sakhalin, the poorest, most desperate region on the island. There was something symbolic and moving in this development, for these time-honored soldiers were fierce, and on Sakhalin they lived by a code of ethics that represented some of Russia's deepest traditions. Cossacks here rode horses and fought with swords and whips. They lived above the fray, and, unlike most everything else in Russia, they were not for sale.

Guido urged me to see what was happening on Sakhalin Island before they closed the doors completely. I could get in on a tourist visa; the Russians shouldn't have a reason to detain me. The time to go would be early spring, when the salmon started coming in from the sea and when the poachers set up their deadly camps. The Russians of the Far East were isolated, but they were strong. The fight for the salmon would come down to them.

I flew to Sakhalin Island in April 2014. My plan was to visit the watershed council of Uglegorsk, a city in the north that had suffered the worst environmental abuses in the Russian Far East—and where the Cossacks were starting to turn out in numbers. The stories of environmental abuse reached me before I arrived. On the plane from Seoul to Sakhalin, my seatmate, an oil engineer, related quick, brutal anecdotes about the hushed-up reality of Sakhalin's oil industry. Russians had been drilling for oil in a nuclear waste zone. He reported that when asked about the dangers of such a practice, the Russian energy minister had replied with a laugh, "A little radiation never hurt anyone." And then there was his friend, a European engineer working on a Russian floating rig who, in the dead of night, caught the operations manager dumping contaminated concrete into the Sea

of Okhotsk. It was easier and cheaper than disposing of it the proper
way. When he caught him the second time, the engineer told the manager he would report him. The Russian replied, "Fine. Report me.
But know that I have a wife and kids, and if I lose my job I will hunt
you down and kill you."

My seatmate stared into his Coke. "There's an unbelievable
amount of oil here, and there's no regulation at all. You can buy or
kill your way out of any trouble."

As we approached Sakhalin Island's brown, flat, wet terrain, we
could see, beyond the low clouds, the specter of mountains in the
distance. The runway was in ill repair, and we jostled and jerked to
an uncertain halt, and tentative applause. Ours was the only plane on
the tarmac. A few officials wandered out from a monolithic gray
structure as if awakened from a nap. They wore stiff green military
hats with outsized circumferences and stood in an uneven line waiting for the stairs to be pushed across the cracked tarmac to the plane
door.

Driving into the city of Uzhno-Sakhalinsk, I tried to absorb the
poverty of the island. The roads were appalling. The city looked
bombed out; buildings were half-finished and crumbling. In the lobby
of my hotel, a television blared with state-sponsored coverage from
the Ukraine and Syria that featured many heroic Russians. I retreated
to my room, which was tidy but cold. I pulled on socks and a sweater
and looked out the window, where there was a pile of junk, broken-
down sheds, and weed-covered abandoned cars.

That night I met with members of the local watershed council in
the hotel restaurant. The warm, animated group spoke in whispers,
as if the room was bugged. Guido had told me about Maxim Ageev,
a handsome young schoolteacher, who had become the Wild Salmon
Center's local hero. Maxim described what he had seen on the Lu-
toga, his home river. Two hundred miles into the untracked wilderness, he had come upon a poaching camp from the trees. Rifles leaned
against the riverside tents, and a portable television blared from a
canvas tabletop littered with satellite phones, cigarettes, and radios.
Some distance away was an armored personnel carrier covered with

a tarp. Maxim counted five men, decked out in expensive camouflage. Two of the men were sitting and smoking. Three were standing
in the river surrounded by the chrome backs of hundreds of salmon.
Stretched across the river was a net. Maxim snapped another shot of
the poachers plucking the fish from the water and slitting open their
bellies. On the riverbank and spilling into the forest was a growing
mound of gutted female salmon. Thrown alongside, the males suffo-
cated slowly. Buried in the soft soil were casks filled with salted roe.
Days or weeks later, the casks would be hauled out by helicopter and
sold for millions in Korea, China, and Japan.

Maxim, resolved to fight a nonviolent war, had emerged from the
woods and greeted the poachers pleasantly. "Did you know that
poaching is illegal?" he asked them. The men watched wordlessly as
Maxim walked to the net and cut it loose. Released from its hold, the
net swung downstream and the fish started to move in a glinting silvery mass. Some, trapped in their watery corral, could not get free.
Others headed downstream, back to the ocean. Only a few continued
upriver. Maxim distributed some pamphlets to the poachers. "Here,
you can read more about it."

It was dangerous work, but what choice did he have? He wanted
his children to know what salmon were, and his peaceful efforts were
having an effect. Besides, he was no longer alone. This past year,
Maxim led a passionate platoon of locals to conduct 110 raids on ten
rivers for over twelve thousand miles, initiating twelve administrative
proceedings and fifteen criminal ones. He was sure the tide was turning. Corrupt officials were considering the law again. People were
paying attention. In the north, it was harder.

The next morning I was joined by two companions from the local
watershed council and Roman Shatrov, a charismatic young conservationist from Sakhalin Environmental Watch. We climbed into a
rugged-looking vehicle and followed the only road north. Uglegorsk,
which translates simply into "coal town," lay nine hours north of the
capital. I was told that Uglegorsk's once-thriving economy had included robust salmon and crab fisheries. Since perestroika, the fish-
processing factories lay dormant. Now the surrounding hills were

gouged for coal that was shipped to China. I learned that only the previous year did the coal trucks stop driving through the center of town, blackening the streets and buildings. The cancer rate was high in Uglegorsk; young people were dying of cancer of all kinds: breast, brain, stomach, lung. Instead of accepting their unjust sentence, the people of Uglegorsk were making moves to reclaim their land and water. They had recently elected a mayor whose priority was to stop the unchecked environmental abuses and nurse the rivers and hills back to health.

The day was drizzly and gray and the wind hit us hard as it charged across the landscape. Roman knew the names of every plant, flower, and tree, and, as we passed boreal forests, he told me pointedly that some of these forests were transitional; the evergreen trees that grew in the darkness below the birches were small and ugly, but when they got enough light they shot up—and when they were big and tall they overshadowed the birches—and the birches died. It was the way with natural succession, and I understood that this was a metaphor for what was happening on Sakhalin.

We pulled over at a desolate beach with a whipping wind and a few huddled gulls. Here it was 33 degrees and we were headed to the mountains. Once over the crest of the pass, the 100,000 rivers and creeks that flowed east started flowing west. Winding down we reached the western coast where the water was calm and the air warmer. There were strange sights: abandoned Japanese structures in the middle of empty fields; cows standing placidly in the sea, taking "sea baths."

We stopped at a cluster of small flat-roofed structures that was not so much a village as a supply station. The outhouse, a pit in the ground, cost fifty rubles to use. A shop selling basic provisions was tended to by a stolid woman who sat behind the counter with a silky long-haired cat. I pet the cat. "She's beautiful," I told the unsmiling shop owner. "She bites," the woman said. Outside, a pack of wild dogs roamed close to us, looking for handouts. The wind fluttered plastic bags caught in the bare branches of shrubs and dwarf trees.

A mud-splattered sign announced our entrance into the Uglegorsk region, and immediately the brown road turned to black with spilt coal. We wound down to a wide, fertile valley where in the last century the Japanese grew rice, fruit, and vegetables. The valley had lain fallow since the Russians had forced them to leave after the Second World War.

We pulled over to watch a coal fire burning in an empty field. The ground was so saturated with the combustible rock that the flames hopped like rabbits as the fire spread. Coal trucks passed us traveling at top speeds. I counted one truck per minute. The coal was heaped high and uncovered and was on its way from the hills to the port of Uglegorsk. Overloaded, the trucks tipped over on a regular basis, spilling the coal and often killing the driver. The fire burned untended in the fields, blazing as bright as Hades, the smoke billowing in pitch-black columns.

I was secretly dreading Uglegorsk. Once, this city had been in competition with Uzhno to be the capital of Sakhalin. Now, I was told, five-year-olds smoked cigarettes on the streets and patrols of wild dogs made some neighborhoods impassable. In the city, a few bony cows wandered through the central square, which was not actually square at all but a large misshapen patch of broken road and dirt. It was a one-street town surrounded by an industry that gave them few jobs and no income.

The hotel was on a side street and didn't look like much, but I was told it was Mafia owned and so the best in town. At the entrance there was a bucket of blue paper booties to slip over shoes to keep coal dust from smudging the clean white floors. My room was embarrassingly luxurious, like something from Vegas—a suite with mirrors and chandeliers and two huge flat-screen televisions. I had oversized windows that faced directly into an old office building that looked like it had been bombed.

I was advised to take a sauna, the Russian panacea for all ills, particularly after nine-hour jouncing car rides. The hotel's *banya* facility was also Vegas-style, with gilded tiles and statues of deities. I was given an elfin wool hat to protect my hair, and for an hour I rotated from the cold pool to the steam sauna to the dry-heat sauna, where I lay on the hot cedar planks and considered the strangeness of where I was. Afterward all I wanted to do was sleep, but the main event was yet to come, at midnight, when the local Cossack chief was going to take us night fishing.

Pavel was tall and broad and solid as a tree. His face was clear and handsome, and his blue eyes were full of merriment and curiosity as he grasped my hand in his. He was glad I had come, he said. They all were. It was an auspicious moment, after all—just a day earlier, the capelin had arrived on the western beaches of the island. It had been fourteen years since the fish had appeared in such numbers. No one knew why they were gone so long or why they were coming back now. Some thought an earthquake in the Kuril Islands had brought the fish back to Sakhalin; some thought the ocean spirits were favoring them again. Either way, the people of Uglegorsk considered it a good omen. I was learning that this part of the world was full of metaphor and symbolism, that nothing was without meaning, especially in nature.

Pavel and his girlfriend, Anna, drove our little group across town to the beach, where small fires burned and clusters of people gathered around buckets. In the moonlight, I could see thousands of the small silver fish wriggling in the sand as the waves delivered them during their frenzied nocturnal spawning. People up and down the beach waded into the sea with nets and scooped the fish into buckets. Later, Anna told me, they would be delicately gutted and fried or dried. She would stay up all night helping. We watched as Pavel strode into the breaking waves like an oceanic demigod and thrust his homemade net into the surf. He returned to his frozen womenfolk as we huddled around the rising bucket of squirming, expiring fish.

When they dropped me off at the hotel, Pavel told me that the next day we would go for a traditional Cossack shish kebab with the members of the local watershed council. This was where I could meet the team. When I asked where the shish kebab would be, Pavel answered cheerfully, "In nature."

Pavel's SUV broke down when we were well and truly in nature. We were halfway through a mountain range after driving miles along the coast, past abandoned fishery structures. A few ramshackle homes advertised crab for sale by placing a crab shell on a chair along the road. It was impossible to tell from the outside which houses were inhabited and which lay empty. They all had the look of desertion.

It wasn't a great day for an outdoor meal. We waited in the gray drizzle while Anna pulled out the platter of capelin she had spent the night preparing, and we ate them for lunch in the car. The fish were fried in light batter and were delicious. The tiny bones piled up in the bag at our feet. A few cars passed and stopped, recognizing Pavel. As a Cossack leader, he was known and respected by Cossacks and non-Cossacks alike. I learned that he was currently staging a land and horse acquisition for his community. They had registered their plot of land for farming and bought cows and pigs and chickens. They were going to build a small paradise. Cossacks here, he said, were unified by an ethos rooted in justice. To live within the community, they had to understand and adhere to Cossack law, which included working the land and developing their spirituality. It also designated them as protectors of nature. Conservation of salmon was written into their bylaws. Their new horses, Pavel told me, would help them drive out the poachers.

We sat on the muddy road and ate capelin until we were rescued by our shish-kebab mates, who drove us over the rutted mountain roads to the mouth of the Pokosnaya River. A year earlier, Pavel had discovered poachers at its mouth. These armed men had netted the width of the river and on the banks were piles of gutted fish three feet high. It made him cry, he said. This was the moment he committed himself to fighting the poachers.

Surrounded by boreal forests and thick patches of snow, we set up the shish kebab at the river's mouth. The other members of the council eyed me shyly and with a tinge of wariness; I was an outsider, and an American no less. Was I there to spy on them? Watershed councils were not supported by everyone. The Mafia was powerful on Sakhalin, and they were heavily invested in the poaching industry. Council members kept a low profile and did not speak openly about their plans. Such civic activism was frowned upon in Russia. They had to tread very carefully, eluding the attention of both the Mafia and the FSB. They had allowed me to join their ranks because I was a writer, and because I was Guido Rahr's first cousin. In Russia this was the same as a sister. They knew of Guido and all the Wild Salmon Center had done for Russia's salmon. They had heard he was a pure soul, and deeply connected to nature. I could sense a current of excitement in them as they watched me, as if there were things they were eager to share.

The rain had stopped, but the day was cold and windless, and the sea glassy calm. Pavel pulled out a rustic homemade grill from the trunk of his car and we started gathering firewood. There wasn't much to be had. I took a long walk down the beach and returned with a meager offering of kindling. The others had clustered in groups and chatted as they leisurely prepared the meal. The wind had kicked up and it was now freezing.

Meanwhile, the Russians were in heaven. For these people, Russia was its land: its flora and fauna; its berries, nuts, and mushrooms; its livestock, vegetables, and fruit. The Cossacks didn't view themselves as separate from nature; they honored its offerings and fought for its protection. The offense against their salmon was like an attack on Russia itself.

Soon the vodka was flowing. The guitar came out and there were songs beautifully sung. We toasted to Mother Nature and salmon, and I tried to ignore the cold. These people were seriously tough, and I got the feeling the party was just getting started. When we ran out of wood, the Cossacks marched into the forest with their axes and felled an old tree. With a new infusion of fuel, the cooking fire burned brighter, but wasn't nearly warm enough.

By this point, nature and vodka had purified us, and we were friends. I met two teachers, an investigator, and a fisherman. They spoke passionately about their mission to stop the poaching and to teach the people of Sakhalin the value of salmon. I was doing my best to follow the conversation, cranking the rusty wheels of a language I knew twenty years ago. The group had let me in the way Russians can—wholeheartedly. We toasted to friendship, and the team. The vodka was making me warm inside. The small folding table was set like a rustic still life, with capelin, shish-kebab pork, and vegetables. The vegetables were served in their natural state; only cucumbers and tomatoes were cut into rough chunks. The Cossacks had gathered herbs and wild garlic and scallions, and we ate these on their stalks. Along with the creeping euphoria I was feeling, I was also aware of a slight change in key, as if the tone of the conversation was shifting from major to minor. I realized it was time for us to talk—to really talk.

Vasily, an older and fiercer Cossack chief with the countenance of a war god, wanted to know my thoughts on Ukraine. I was tipsy enough to tell the truncated and diplomatic truth, as I understood it. While Ukraine and Russia were deeply connected, the birthrate in the region hadn't favored Russians, and the majority of the Ukrainian population was leaning away from Russia both economically and culturally. It was a difficult situation. I added that I understood that Ukraine, with its port and pipeline, was strategically important for Moscow, and I could understand why Putin was fighting for it.

The Russians listened in silence, and when I finished, Vasily and his Greek chorus delivered their rebuttal like a rapid-fire offensive. First, Ukraine was a part of Russia; it always had been. In any case, had I known Russia to go beyond its empire to conquer or occupy? Was it greedy for the world's riches? No! Had it stolen from other countries? No! Russia wanted only to live in peace. Pavel spread his arms and declared that Russia was a great and big country, half in the west, half in the east; there was a saying that the rebirth of the world would come from here. "We have always lived with nature," he said, "and nature is god—and god is love. We are here to love each other. This is the Cossack way."

"It is our clarity that makes us strong," Vasily added. "We know what we're fighting for. There is nothing stronger than a Russian soldier, because he knows what he's fighting for."

When the offensive was over, the group stared at the fire and collapsed into a melancholic torpor. The Cossacks were genuinely sad about what was happening in the United States over Ukraine. "Because we are all one. To create a war is wrong. It is a mistake. They are feeding lies to the West. There are things the Western media doesn't know. You don't have the full story."

"All Russians want is to be left alone with their land," Vasily echoed.

"Then why don't they take better care of it?" I asked, tired of staying quiet. I had inside information, after all. "There is nuclear waste, contaminated oil dumping, coal toxicity. Is this love?"

"The problem," Vasily said, "is with upper management. The problem is with the oligarchs. They are creating the trouble."

Everyone nodded at this. I told them it was the same in America; we also had trouble with the upper management. But a country is not its upper management, I suggested. "No!" the Russians agreed. "A country is not its upper management!" There was another toast to this unexpected common ground.

Vasily had taken control of my care, making sure I had enough to eat and drink. Seeing that I was cold, he led me to his van, where I could warm up. I sat in the driver's seat, next to his knife and Cossack whip. The van had been jerry-rigged to go anywhere, with fat tires and the clearance of an eighteen-wheeler.

We left as it got dark. Vasily had a good amount of vodka on board and tore along the road with one hand on the wheel and the other searching for a Cossack CD he wanted me to hear. The seatbelts didn't work, and I hung on to the door as I bounced around in the passenger seat. Vasily found the CD and shoved it into the stereo. An accordion blared over the sound system and the Cossacks in the back broke into song. I saw that Vasily was driving 70 miles an hour in a night so dark it was hard to see the road. "This is a song about nature," he shouted. "Cossacks and nature!" In that moment, a white hare bounded across the road, flashing briefly in our headlights. Vasily exclaimed loudly to the men behind him, who gave a big whoop. This, he explained, was a good omen. He skipped to his next-favorite song. It was about Cossack soldiers and had a thrumming, military drumbeat and a catchy tune. I watched as another white hare dashed into the road and the Cossacks exclaimed with more joy. Their hands were in the air. No one sees two rabbits in one night!

Stars were piercing the indigo sky, and I was thinking that there were worse ways to die when the third hare bounded across the road. At this, the Cossacks dissolved. A prophecy was declared. A few moments later, Vasily stopped the van and turned off the engine. It was suddenly very quiet. The other cars were far behind us. In front of us the moon hung over the Sea of Japan, laying a silver path across the distant water.

"Come," Vasily said. "It's time for you to shoot a gun."

Sure, I thought. Why not?

The other Cossacks piled out of the van to watch. Vasily reached for his gun, as serious as a chieftain on the eve of battle. I was picturing a nineteenth-century engraved heirloom that had been passed down for generations, but he handed me a Makarov handgun, the kind Russian police used. "Shoot it," he said. The gun felt strangely light. I wondered if it was real. I aimed it at the moon and fired. The night before me burst bright orange, and for an instant there was no sea or moon and I couldn't hear a thing. How frightening, I thought. How easy. Just as suddenly it was over and we piled back into the van and continued the hell ride back to Uglegorsk, my ears ringing, my body jolted to the core.

At the Mafia hotel we were joined by the others and we headed downstairs to the Vegas *banya*. Wrapped in towels and the woolen elf hats, we sweated out the vodka together. A bound bushel of oak leaves was extracted, dripping, from a wooden bucket of water and used to beat our exposed parts. This was good for the skin, I was told, good for circulation, and, of course, good for general well-being.

In the next days our merry band toured more rivers and villages. Away from the coal-blackened city, the life of the island unfurled before us. We saw flocks of migrating swans, seal pups, foxes. I was taken to see the Cossack Valhalla project, where the Cossacks were going to build their utopian society. We built a fire on the grassy field above the sea and made a shish kebab, and I listened to their plans.

Later in the day, the talk turned to the salmon forum, which all the city had been preparing for. Presiding over it would be their new mayor, the first democratically elected mayor on the island. He was special, this man. He understood the earth. Even the Cossacks spoke of him with reverence.

The next morning it snowed, and we walked to city hall, which sat on a hill in the middle of the city. The spacious hall was nearly full, and a hush fell as the Cossack leaders filed in, handsomely dressed in traditional uniform. I noticed that the periphery of the room was filled with displays. Children had drawn diagrams describing a salmon's life and the different species of salmon that swam in Sakhalin's rivers.

The program began with a few highly choreographed, synchronized dance numbers and some soaringly ardent ballads by a group of talented and well-trained children. Then the mayor took the lectern. He was dashing in a well-cut suit. In one hand he held up an environmental report that showed that Uglegorsk was leading the race in environmental disasters in Russia. It was shameful, he said. This was their call to action! He was charismatic with a wiry intensity. Listening to him, the audience was slightly restive, as if they were not yet entirely comfortable with their participation in this display of civil society.

The mayor was speaking about their successes in the past year of combined efforts to fight coal companies for clean water, repaired roads, and the protection of their mountains and rivers. He was angry about the coal that earned 100 million rubles a year and destroyed the roads, the water, the air—and where did the coal go? To China, Japan, and Korea. Where did the profit go? Out of the region. Russia, he said, was looking more and more like America—outsourcing their dirty work. There was a double standard in the world's environmental system. Protection on the one hand, abuse on the other.

Western countries, the mayor concluded, were worried about Russia becoming too strong. The West was trying to weaken Russia with its Ukraine politics—all while the climate was changing, drying up the rain, raising the temperatures. These were big problems, and they belonged to all of us. We had to create a sustainable, regulated system for the coal companies. We had lots of problems, he concluded, local and global—and we needed to get to work. Good luck to us all! He slapped his dossier closed and marched off the stage to enthusiastic applause.

The forum ended with reports detailing how rivers had been cleaned up; educational programs and camps set up for children, troubled youth, and families; grants won, work coordinated with the police and district government. The council wanted to participate in an international forum where they could present their research and work, highlighting how they had introduced new sectors of the community to antipoaching activities.

The forum was a success, maybe too much of a success. The next day our merry van was replaced with two black SUVs with drivers in black suits, and our group was divided in two. My raucous friends were quiet and watchful and adopted a polite formality as we set out on a tour of Japanese temple ruins and some of Sakhalin's rivers. This was the FSB, and I was clearly the person of interest. There were whispered warnings from my friends not to talk of politics, or even mention Putin's name. I played the ignorant tourist. Every photo I took was examined by our driver. "Why you take this photo?" he asked. "Because it's a beautiful river," I said, showing him the shot on my phone screen. "Don't you think?" The photo was taken in the soft light of late afternoon from an old wooden bridge spanning a peaceful river bordered by wintery birches. The man examined the photo and then nodded. "Yes, it's beautiful."

I continued to chat mindlessly about the beauty of Sakhalin's rivers as we drove on. Outside, the sky was blue as we passed sunlit fields and birch forests. The mountains were covered in mist, and the chimneys of falling-down houses smoked and were somehow cheerful. It was the brief moment when winter danced with spring. There were intermittent snow flurries, while on the ground were clusters of blue flowers and bright green elephant's ears.

For our last dinner we went to a Chinese restaurant. My comrades spoke of other parts of Sakhalin, where people had to poach just to survive. "Even I poached when I was young," a young woman named Tanya told me. "I had no idea what a salmon was until one day I saw one fish swimming one way and a bigger fish swimming the other way." Tanya agreed that most of the population knew nothing about salmon. "We have been developing salmon education programs, which is how we've learned about them. Ten districts participate in regional salmon festivals and forums. This is how people will learn."

The group went silent, as if they knew somehow these festivals and forums could never be enough.

"If we still had the Soviet collective farms, we'd have everything," Vasily said, to general nods. "We grew so many things—we even had a pig farm."

Someone else chimed in, "We had cottage cheese and sour cream that we provided for the whole island. But when democracy came it all came crumbling down."

"Now all people think about is commerce," Tanya said mournfully. "I don't think this free market has led to freedom. There is so much in the hands of so few."

But why dwell on such things? Tanya said. To change things, one had to embrace hope, not harp on the past or grow bitter over present helplessness. The laws of nature were at work everywhere; it was only a matter of time. The old growth would die away, just as in the birch forest. The future was with the children, and the children were learning.

Only months after I left my new friends, the Wild Salmon Center lost its last employee on the ground in Russia. Guido and his team had realized by then that supporting their local Russian partners was the most effective way to move forward. They focused on becoming "the ultimate ally," offering these isolated groups funding and science, and connecting them to a larger network of environmentalists. Over time these partners would strengthen and grow and become exemplary protectors of their valuable watersheds. Some partners achieved global recognition for their courageous work. Dmitry Lisitsyn of Sakhalin Environmental Watch was awarded the prestigious

Goldman Prize for protecting Sakhalin Island's critically endangered ecosystems, including some of the island's salmon rivers. On the mainland, Alexander Kulikov, a wildlife biologist and president of the Khabarovsk Wildlife Foundation, protected 1.6 million acres of pristine salmon and taimen habitat from logging, mining, and road construction. These, Guido maintained, were the true heroes—the Wild Salmon Center could offer only a supporting role.

CHAPTER 28

THE APEX PREDATORS

UNLIKE HIS ORGANIZATION, Guido did not retract from Russia. There were still mysteries to uncover, and one in particular held him rapt. By 2015 he had become obsessed with the Tugur River. He had now learned that the rivers to the north and south of the Tugur had been seriously degraded by poachers. These rivers, the Uda and the Nimelin, had once held giant taimen too, but the rumor was that the majority of these fish had been netted and killed, while caviar poachers had diminished the salmon populations. As far as Guido knew, the Tugur was the last pristine giant taimen river in the world, and he was on the verge of begging Alexander Abramov to let him see it again. It wasn't just his lust for catching the gigantic fish; it was to ascertain their essential nature. DNA analysis of Sakhalin taimen, a sea-run species, had shown that they were the most ancient salmonid on earth. The taimen of the Tugur were the biggest of these dinosaur fish, but next to nothing was known about them. Without a genetic analysis, ichthyologists couldn't even guess at the basics—like how to tell a male from a female. They didn't know population size, or where the fish traveled in the river. They didn't know if the taimen were resident to certain areas, or if they roamed. It was possible that they migrated during the year and, like Sakhalin taimen, went to the sea.

The Tugur itself was equally mysterious. A satellite photo of the river's newly protected area showed mile after mile of impenetrable forest. This fifty-mile stretch of river was impossible to see. Few, if any, had explored it—and it was unlikely anyone had cast a fly on its waters. And Abramov, a man who seemed to be the Tugur's perfect keeper, was every bit as impenetrable.

Guido talked about the Tugur constantly. The more I heard, the more chilled I was by a river and a fish that seemed as unknowable and dangerous as Russia itself. Guido felt none of my uneasiness as he marshaled every resource he had to penetrate what seemed to be the country's heart of darkness.

Guido called me in the spring of 2015 and told me that after months of email overtures, Abramov had finally invited him to the Tugur. It would be a private trip, with close friends and associates for a week in early October. They would fish for taimen, and Guido could show them what he could do with a fly rod. Afterward, Abramov approved an expedition for Guido to travel through the new protected area with a small group of elite fly fishermen who specialized in catching taimen. They would take their rods and spend over a week floating fifty miles down part of the river that was wild and unknown, even to their host. For the remainder of the spring and summer, Guido thought of little else, delving into whatever material he could find on taimen, and experimenting with new flies. For him, the Tugur had come to represent a new category of stronghold, one that had yet to be fathomed by modern man.

In August, Guido called to tell me the trip was set, and that there might be one spot left if I wanted to go. A trill of excitement ran through me. If I could get myself to Khabarovsk, Misha and the other

anglers would meet me there and we would make our way to the
Tugur. Guido would join us there directly from Abramov's lodge. It
wouldn't be an easy journey, he added.

It seems impossible that it takes three days to get anywhere in the
world, but in mid-October, flights to parts of Russia slow down, or
stop altogether. I made my halting way east and then north, flying
from Seoul to Vladivostok and then to Khabarovsk, watching as the
human density of Asia gave way to the arboreal paradise of eastern
Russia. Deciduous forests in peak autumn color carpeted the hills and
mountains that were interrupted briefly when we landed in Vladivostok, at a sleepy airport with abandoned Soviet planes painted with
faded red stars barely visible through the weeds that had grown up
around them. It spoke to the vastness of the Russian empire, of dis-
carded regimes and weaponry and so much space that there was no
need to bury the remains.

During my brief stay in Vladivostok, I was terrorized by bureaucratic dysfunction. My plane reservation did not exist. Then it did. I
was ordered to condense my two bags into one, or else leave one bag
behind. I broke into a sweat repacking my bulky gear in the stifling
and airless airport. When my luggage was finally checked I stared
after it, realizing I might not see it again. I made my way outside to
breathe the fresh air, passing tchotchke shops and a café. A fish shop
featured a long glass case displaying rows upon rows of plastic containers filled with salmon caviar.

We flew north at sunset to Khabarovsk, passing more colorful forests, muted by the rose hue of the sky. There were rivers below, one
after another, snaking and glinting across the land like silver ribbons.
We landed in darkness. Misha picked me up at the airport; I knew
him instantly. His quick gray-green eyes knew me too. We talked as
he drove me in his beat-up four-wheeler to my hotel. I had the distinct
feeling that I had met my cousin's Russian counterpart, and we fell
into an easy companionship. Guido had sent him a shopping list for
the trip, and the next day we went to a Russian megastore to buy

vodka, beer, sausage, cheese, chocolate, bread, and water before pick-
ing up the rest of the expedition, three world-class fly fishermen who
had flown in from Japan, Mongolia, and Georgia. We were told it
would be a fifteen-hour bus ride north.

The plan was to arrive in the town of Briakan in the morning, and
from there chopper to the Tugur. The five of us settled into the bus
and watched the lights of the city fade away. Soon we passed out of
cell range and were on an unpaved road traveling north in darkness.
It was too bumpy to sleep, and we bounced and jostled along the pitted road that was slow with mud and rain from a typhoon that had
roared up from Japan and wailed outside in the darkness.

At three A.M. the bus broke down. The driver got out and made
some discouraging banging sounds under the carriage. We sat for one
hour, and then another. At five A.M. he turned off the engine and announced we would be spending the night on the road. The problem
with the bus required the light of day. A cold front was moving in; we
could all feel it. There was nothing to do but put on more layers of
clothes and wait for the gradual lightening of the dome of clouds
overhead. At 7:30 A.M. the driver got back under the bus with his
tools, and by 8:10 the bus was miraculously lurching to life.

When we finally arrived in Briakan, our group was cold, hungry,
and unsettled. We had missed our window to fly. The clouds had
settled close to the hills, and the Mi-8 helicopter sat idle on the flats
outside the town. The pilots were nowhere to be found. We stood
staring at the lifeless chopper as rain pelted us and stray dogs sniffed
us for food. With nowhere to go, we headed into Briakan to find
some warmth. The only place open to us was the general store, a tiny
shack sitting among other shacks in a muddy clearing. We pushed
into a crowded space that was stacked from floor to ceiling with
goods, and a rotund, elderly woman added up the price of our chips
and apples on a big wooden abacus with round beads. "Russian computer," she murmured. The café next door was empty; there was no
coffee and no tea. When one of our members found a wall outlet to
plug in his phone, there was a bang like a gunshot and the lights went
out, not just in the café but in all of Briakan.

We would not be flying that day, Misha told us, and the electricity in Briakan would not be coming on anytime soon. We would have to make our way to the next town to spend the night. We boarded the bus and settled in to ride for a few more hours. My angler companions lapsed in and out of forced cheer, talking fish to keep themselves going: leaders, tippets, casts, presentation—layers upon layers of intricate considerations, all of which brought them briefly to life as they recalled the many great fish they had fought. Between them, it seemed they had been everywhere. Theirs was a world that existed in terms not of countries but of rivers. What they did not speak about was the possibility that, with so much rain, the Tugur would be blown out. A river with too much water turned muddy and the fish were hard, if not impossible, to find.

We booked into the only hotel on the one-street town of Polina Osipenko, a town named for the heroic Soviet female pilot who crash-landed her plane in a nearby forest. The mattress in my room was an inch thick and hurt to sit on. In the bathroom was a tub barely big enough for a small child, with water that was tepid and brown with rust. The rain continued all night, drumming on a corrugated roof outside my window. As water dripped onto my pillow from one of the three leaks above my bed, a man I had seen earlier at the hotel bar banged on my door hourly, drunk and belligerent. I put in my earplugs and wondered how Guido was faring at Abramov's luxury lodge.

We woke at five A.M. for the drive back to Briakan, to see if the chopper pilots would consider flying. At eight A.M. they appeared like a SWAT team—three of them with cigarettes, dark glasses, and leather jackets, yelling at us like we were cattle: "Go go go! Time is money!" They rushed our bus down to the helicopter and threw our gear from the holding compartment into the mud. We carried boxes and boxes of food and our oversized duffels of gear into the chopper, which was gutted and voluminous inside. As we clambered on board, our pilots scolded us for being slow, and then pointed to the headphones as they fired up the engine and turned their backs to us. Even muffled by headphones, the noise was tremendous, and as it crescendoed the

Mi-8 started to rock back and forth, lifting off the ground and then touching back down. When the rotors kicked in to full intensity, we rose slowly into the air. Our weariness fell away entirely as we skimmed yellow birch forests and hills and flew low over bogs and tributaries.

An hour later we passed over a braid of streams that led to bigger streams, which led to minor rivers that curved and bent and parted around fallen trees and logjams. Then we were directly above the confluence where all this water connected and became one huge river. The Tugur. I strained to see as much as I could from my little window. The river was immense and wound around and through forests. It was impossible to see how big it was; even far to the east glimmers of water wove through distant trees. Along the banks was the uniform debris of logs, entire forests that had been uprooted by the water's changing course. The river had shorn them of leaves and branches as they rolled and bumped up against each other, sometimes lining up neatly, sometimes driving up like swords and clashing in a chaotic tangle of branches and smaller trees. The amount of wood was staggering. It looked as if a timber yard had been spilled onto the river.

We had not been following the Tugur for long before the pilot circled a large gravel bar and began his descent. Two Steller's sea eagles were flushed from their perches, their broad wings extended as they angled down and away from the wind of the chopper. The copilot climbed backward out of the cockpit, bit down on his cigarette, and extended a short ladder from the open door as we hovered above the gravel bar and then touched down. He jumped to the ground and waved us impatiently to follow. There would be no stopping; the weather was changing, and we had to unload as quickly as possible. We threw our gear into a pile on the rocks below, deafened by the din of the hovering chopper. When the Mi-8 lifted off, we lay on top of the bags so that nothing would blow away, shielding our faces from the violent whirl of debris that was kicked up as the chopper rocked, rose, nosed forward, and was gone.

We were suddenly alone in the silence of the wilderness. At some indeterminate point, the chopper would return with our guides, the

few men who knew something about the Tugur and were willing to
float it this late in the year. Until then we were in suspension, staring
at the sky and listening for a far-off thrum. As we waited, propped up
on our luggage, the temperature dropped. I listened as the anglers
discussed the size of the river and calculated how much water was
flowing by, and the force that it would have plowing into a logjam. It
was a tentative assessment of a situation we were just beginning to
grasp, for we would soon be at the mercy of this force. The distant
sound of the Mi-8 returning had us scanning the sky with relief.
When it touched down, another frenetic unloading deposited four
men and a large pile of gear, and then the chopper lifted off, this time
for good.

Our guides were instantly in motion. A small chain saw was dispatched to one of the many beached logs, and soon there were rounds
of wood that served as stools. Other logs were split into kindling with
lightning strikes from an ax that was handled with the artistry of a
knife. Soon a fire was crackling, with a grill laid over it and two cast-
iron pots and a blackened kettle filled with water for tea. The guides
were tough, even for Russians. Oleg Abramov (no relation to Alexander) and his partner, an indigenous Nanai man named Ivan, had a
wordless understanding of each other. I watched as they scouted the
gravel bar for a place to pitch the tents. They had been here just a
week earlier, but the river had dropped a meter since then and the
topography was now unrecognizable. Ours was the last trip they
would take this year. There was concern that it was already too late
in the season.

I wandered down the gravel spit. The Tugur had worn its rocks
into smooth, round pebbles. At the far end of the gravel bar I saw
some bear scat. The brush behind me was thick and obscured everything in its tangle of leaves and branches. I looked back at our little
camp as it rose from a sea of smooth rocks and felt suddenly uncomfortable being so far away. At some point, Guido would join us from
Abramov's lodge, some forty miles upriver. I wondered about the logistics of meeting someone at such a designation. Where would he

come from, and how would he get here? I realized how much I was looking forward to seeing him.

In the dining tent it was warmer, and a long table had been set up with biscuits and dried fruit and salmon roe. After enduring two days of cold, I was finally offered hot tea. The guides seemed indifferent to me and treated me as they treated the men, with businesslike respect. They rarely saw women in the field—and they had never seen fly fishermen. I watched them cast dubious glances at the anglers in our party, who were setting up their fishing gear, piecing together long, flimsy graphite rods. On the Tugur, taimen were fished for with stout spinning rods, a strong, simple reel, treble hooks, and sometimes a gun. Oleg communicated his dour predictions for the coming week. He did not believe this fly-fishing technique would catch much of anything, and certainly not a trophy taimen. And anyway, he said, the river was very high. The anglers should be prepared to catch nothing at all. Our group accepted this dismal forecast stoically and dispersed, heading up and down the river. All three fishermen were soon out of sight.

I was sitting making notes in the dining tent when one of our party came back from the river. Mark was an expert taimen fisherman who had joined the expedition from Japan. He had been preparing for the Tugur for months, rigging up buckets of water on a pulley in his small Tokyo apartment to test the strength of his leader. Mark was good natured and had weathered the vicissitudes of our journey with humor. But he was looking decidedly pale now as he stood in his fishing gear and related what he had just seen. He had been wading down a deep riffle when his foot had brushed something. He thought it was a branch, but when he looked down, he saw the biggest fish he'd ever seen. It was at least five feet long, and it was right there, at his feet. He stopped and watched as a taimen swam slowly past, its green bulk followed by a huge red tail that swished back and forth as it disappeared into the darkness. Then, a moment later, the fish turned and faced him. "It looked right at me," he said. "I've never seen that." And then he went silent. Mark was done fishing for the day, he said.

The fishing gods had given him a sign, but it was unclear what that sign meant.

In the early afternoon we heard the distant hum of engines. A moment later two high-powered riverboats swept around the bend, slicing through the river like knives. The gleaming boats pulled up to the gravel bar and dispatched six men dressed in camouflage, two of them armed with AK-47s—Abramov's security detail. We watched, speechless, as the men with guns scouted the gravel bar and took up guard positions, facing opposite directions. Guido was let out of the boat last, and he threw his hands in the air in greeting. He seemed very happy to see us.

The rest of Abramov's detail smoked cigarettes near the boats and chatted with Oleg. A guard with an AK-47 wandered toward me to have a look at our gear. His black beanie was pulled low over his shaved head, and he motioned me over as he inspected the rafts. He was a mountain of a man, with a full beard and the broad features of people from the steppe. He pointed to the grommets at the bottom of the boat and told me the raft was no good, that it would sink. Then he smiled with a mouthful of gold teeth as if to wish me bon voyage.

I looked for my cousin, who had drifted away from the others and was standing on the riverbank, reviving the way a fish revives after it has been released. Looking at him next to the Tugur, he didn't seem big enough for the river. None of us did. I watched him for a while before joining him. When he noticed me, he pulled himself out of contemplation and smiled. "The water clarity isn't bad," he said brightly. "Not bad at all." As we walked back to the others, he said quietly, "We almost got dragged into a logjam on the way here. We came close to losing the motor." Ten miles upstream, a logjam had blocked the entire river. Their jet motor had sucked in some rocks and lost power. They'd had to get to shore to clear the rocks, and then Abramov's men cut a narrow passage through the logs with chain saws. These guys were impressive, Guido said. He forced a laugh as we watched them stamp out their cigarettes, reboard their boats, and push off from the shore. A moment later they were gone, and we stood in silence watching their wakes disappear.

Dark clouds had gathered when Guido suggested we find Misha and get onto the river. Cold rain angled into our eyes as we headed away from camp in Misha's little outboard raft. In the solitude of the river, Guido started to talk; it was the beginning of a long unwinding. The past week had been stressful in ways he had not expected. Abramov had brought his lawyer and his doctor and was never without his bodyguards, even in the sauna. The lodge was comfortable, but it had not been relaxing. While Abramov was a generous and courteous host, flat-screen TVs blared the news from Syria that was vociferously anti-American. Guido had brought his Russian-speaking wingman, Mariusz Wroblewski, who acted as translator. Guido's attempts at chitchat fell flat; innocent American small talk triggered suspicion on Abramov's part. Guido asked too many questions. He had been advised to avoid controversial topics, but he found this hard. In the end, he had broached US-Russian politics, science, and fishing. On each of these subjects, Abramov had been intense and confrontational. He challenged the American's assertions at every opportunity, almost as if to see how he would react. Guido viewed it as some kind of test. A submissive posture would signal weakness, and this, Guido was sure, would end in loss of respect. He elected to hold his ground on everything from the politics of the Middle East to which species of Pacific salmon fought the hardest. Mariusz, who knew Russia and Russians well, approved of Guido's strategy, and the men had heated discussions that went deep into the night. "I argued when I could," Guido told me. "But he beat me almost every round. Honestly, I don't know if I accomplished anything with him." By the end of the week, Guido didn't know if he'd succeeded or failed with Abramov. The Russian gave no indication as to whether he was open to his men working with the Wild Salmon Center, or if this would be their last meeting.

Guido cast long into the current and exhaled. We had found a spot to anchor, and light slanted through the layers of cloud low on the horizon, turning the water silvery. The current spun us in slow circles as Guido balanced himself in the front of Misha's tippy boat and cast one of his hand-tied flies. They were meant to imitate grayling, a small but colorful member of the salmon family. Guido used Icelandic sheep hair to form long wings that would undulate in the current. Their heads were made of spun deer hair. Long chicken hackle feathers and strands of glitter were tied in with the sheep hair to give the fly iridescence and flash to attract the taimen's attention.

We pulled the anchor and putted to a side channel and Guido cast from the boat in the driving rain. "Taimen don't like rain," Misha said. He was hunched and unreadable at the back of the boat, with his hood pulled down low over his face. "Really?" Guido answered, giving no indication that he was going to stop. His two-handed spey rod whipped through the air, forming figure eights, the cumbersome fly slapping the water on either end before it was snapped back, redirected, and hurled across the river again. He fished the hole once and then again, and then reeled in his line, announcing that while he had not had a bite, he was confident that he'd fished the spot well. The question was where the fish were in the river. The water was a meter higher than it normally was. So how to get the fly low enough to tempt a fish? Guido wondered aloud if his fly was swimming too deep, or not deep enough. He asked Misha what he thought, but Misha only shrugged.

Guido continued to talk about Abramov. Later, I would meet Abramov and see for myself this new species of Russian man Guido was grappling with. On a purely physical level, Abramov conveyed the power of an apex predator. Built like a wrestler, he was well over six feet and had the unwavering confidence of a man who had earned his place at the top of the food chain. A PhD in plasma physics, Abramov did not like sloppy thinking, and he did not trust dreamers or idealists. He spoke in harsh realities and had laughed incredulously when Guido told him about the success of the Marine Stewardship Council with "making" Russian fisheries sustainable. He thought it was hogwash that Russian fisheries were paying attention to regulations. Russians didn't pay attention to the law; laws were something to be circumvented. The only real law in Russia was the law of survival. This cynicism was difficult for Guido. "How do you talk to someone who doesn't believe in what you're doing?" he said to me. He added, "But I think Alexander believes in protecting the Tugur." In his voice I heard more hope than conviction.

Abramov did want to protect the Tugur. Like Guido, he had fished since he was a boy. He had been searching for giant taimen for decades, exploring rivers all over Russia, Siberia, and the far east looking for fish that only grandfathers could remember, fairy tales that were now long gone. For decades there had been nothing over fifty pounds caught in their rivers. Still, Abramov kept looking. It was in 2002 that a friend told him about a river more remote than any he had seen.

Abramov found his monster taimen on the Tugur later that year, and they were bigger even than in the fairy tales. By 2003, he had purchased the rights to the river and built a lodge. He spent a month out of every following year studying taimen and figuring out how to catch them. It started with a leader like a steel cable and a lure heavy enough to barrel to the bottom of the river. Abramov found that taimen liked to wait in the shadows of fallen logs and watch what passed by. He discovered the hard way that it was better not to force or fight them to come in once they were hooked. If the fish registered that it was being fought, it would fight back, and break everything from the line to the rod. Unlike most Russians, Abramov learned to treat the taimen gently, and released them always. He wanted them to live.

Bit by bit, Abramov began to understand the taimen of the Tugur. He discovered that the largest fish subsisted on chum salmon when some of them regurgitated three or four chum after being caught. This made sense; the river was full of chum—and it explained the immense size of the taimen. A majority of the chum population spawned in the upper tributaries of the Tugur. Spawning chum had also attracted a number of high-level caviar poachers who came in by helicopter for peak migration weeks. Abramov told Guido that when he'd first arrived, they'd been set up on every major gravel bar. It was thought that the poachers harvested up to fifty tons of chum roe per season. Knowing that such extraction would soon diminish the chum population, and put the primary food source for taimen in jeopardy, Abramov decided the poaching had to stop. It took a team of local fisheries inspectors (unarmed) and a small division of Russian special forces (armed) to drive home the message that the Tugur was no longer open to poachers. The two teams worked together through the high season, hunting poachers with helicopters and boats. It didn't take long for the poachers to understand they were not going to win the small-scale war. It had cost Abramov $45 million to clear the Tugur of poachers. This was how you protected a Russian river, he told Guido, with money and force.

Abramov had gone on to defend the Tugur from successive assaults by loggers, who wanted to build roads and cut down the forests, and by gold miners, who wanted to conduct placer mining at the headwaters of the river, which would be fatal to the fish. Abramov held them at bay with threats of litigation or by purchasing rights himself. For all his worldly success, he believed there was nothing better than being in nature, and he was determined to protect his piece of it.

Misha ferried us to the far side of the river and dropped us off. We started walking, making our way up and over logjams, through bogs. Circumnavigating a small island, we found wolf tracks in the damp soil. Guido paused and studied a back channel that reached deep into the trees, formed a small lake, flowed around an island with new-growth birches, then plowed under a logjam to join the main channel. He took it all in, shaking his head. "I love this river. I could spend the rest of my life studying it."

To do that, he would have to make inroads with Alexander Abramov, who found Guido's proposals to conduct research on the Tugur vague and idealistic. According to Abramov, real research would require tagging thousands of fish and monitoring them with helicopters. Such a project would cost millions of dollars—and Abramov had already spent too much. Guido suggested a smaller project. With a hundred radio transmitters, they could tag and monitor just as many fish. Abramov argued that they might never see these fish again. Guido agreed that was possible, but they had to start somewhere. They should tag one hundred taimen and see what they discovered. At which point Abramov labeled Guido an eternal optimist.

The one scenario that could sway Abramov to invest more in the Tugur had so far proven untenable: a trophy taimen had not been caught on a fly. If the Tugur would qualify as a world-class fly-fishing river, its exposure and its chances for protection would greatly increase. As we made our way back to camp, Guido divulged that Abramov's parting words were, "Catch me a world-record taimen on a fly, and we will have something to talk about."

CHAPTER 29

THE TUGUR

THE CHALLENGE HAD BEEN ISSUED, and Guido was filled with Russia's most dangerous maladaptation: hope. I watched him shake off the darkness of Alexander Abramov as we gathered in the warm dining tent for dinner and were served hot bowls of soup with meat and potatoes. It would be soup for every lunch and dinner from then on, because of the cold. But this was the only night the guides would bother to set up the dining tent.

I was coming to terms with the fact that I had injured my right hand. It had happened in the loading of the chopper, when it had been caught in a strap, twisted, and yanked. But in the numbing cold I had been able to continue using it, zipping and unzipping my waterproof hip pouch to take notes, ignoring that the big zipper was nearly impossible to budge and forcing it over and over again with my thumb and forefinger. This had been a mistake.

The soup and the vodka helped me to forget, as had a retelling of Mark's story. Misha joked that taimen had been known to stalk their prey. Was this giant fish stalking Mark? Mark lapsed into a deep contemplation that we could not shatter that night, or in the days to come. He kept the oversized hood of his arctic jacket closed tight around his face. I had the sense that I never saw him clearly again.

After dinner we swapped tales of the wretched days leading up to this one, which were now hilarious. Guido and I caught each other's eye and laughed. We continued to laugh at things that were only remotely funny. Guido was blowing off steam after his encounter with Abramov; he had not been able to laugh for a week. I was releasing pressure from the three harrowing days it took to get to the middle of nowhere.

Outside, the temperature continued to drop. I got up to relieve myself, wandering away from our settlement until the darkness engulfed me. When I looked back at our bright tent I saw, in silhouette, one of our party toppling backward into the river. The event was slow to register, a capable, solid man flailing in the mud and water. I watched as the Russians pulled him from the river and set him back on his feet.

When I returned to the tent he was standing there, wet, muddy, and seemingly unperturbed by his altered state. He was fine, he said. He needed to get out of his wet clothes, we told him, or he was going to freeze. But he made no move to undress. He insisted he would sleep on the ground; he'd done it many times. The exhaustion and the vodka had made him unreasonable. There was pride there too, and a man not used to losing control. Finally he allowed someone to peel off his wet clothes and get him into his tent and sleeping bag.

The rest of us sat in the tent quietly after he left. We did not need to speak what was abundantly clear. The line between life and death here was thin; one could easily slip away—all it would take was a misstep, a momentary loss of balance. We were vulnerable, every one of us, and we were bound too closely to judge or doubt each other. I wondered if this was how it was with men in war, this unspoken intimacy that rivaled all others.

The tents were not as advertised. Guido had promised stand-up tents, a portable shower, and *banya*—saunas every night. I had brought a towel for this reason. But there would be no washing and no *banya;* the tents were small and thin, with mats laid atop the gravel. I could feel the shape of stones underneath me. I slept badly with the ferocious snoring from the Russians to my right and the Americans to my left, a miserable symphony that even earplugs would not mute. In the middle of the night I was woken by the swish of snow being swept from my tent, once, and then again near dawn.

Daybreak was pale and I lay listening to the sounds of others waking. I could hear from the tone of their voices that winter had come in the night. I stayed in my sleeping bag without stirring, holding my throbbing hand close to my body for warmth. It had swollen up to twice its size. I finally raised myself from the ground with my elbows, and pulled on my many layers unevenly with one hand. The next chores were impossible; I could neither stuff my sleeping bag nor tie my boot laces that were frozen into loopy sculptures. Guido was nowhere to be found. I recruited my new angler friends to help me, feeling like a child. Outside, our gravel bar was white with snow. The eagles across the way were gone.

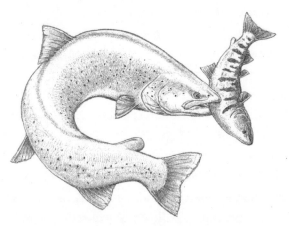

TAIMEN

Over breakfast, our guides shared stories about taimen on the Tugur. There was the monster they had fought for three hours. They never even saw the fish before it broke the hook. Another taimen had banged its jaw on the rocks on the bottom of the river until the lure broke apart and it swam free. Their cargo raft had once bumped up against a fish so big it had raised the end of the raft from the water—

and there was the taimen who had attacked their oars, triggered by their motion into full battle mode.

We would be floating two fishermen to a boat. When it was time to push off, Guido said, "T, you're with me." As when I was a child, I followed my cousin, and for the day was enveloped in the strange peace of his fearless world. Before I got in the raft, I warned him that I had injured my hand and had limited mobility. He seemed to listen, but afterward I wasn't sure he had heard me.

The moment we were swept along in our rafts, we felt our smallness. The Tugur could swallow us without an afterthought. It would take the slightest event, the bump of a floating log or the snag of an underwater branch. A jolt could throw one or more of us out of the boat. If this happened, we had been told, there was little one could do. The cold would chase the breath from our lungs, our waders would fill with water, the currents would tug us downward, and we would be swept along until, like the other branches and logs in the river, we would either be dragged to the bottom or sucked into a log-jam.

We were fishing with Ivan, the native Nanai man. Unlike the other Russian guides, Ivan held my eyes fully upon meeting. His features were broad and handsome, and even in total silence he transmitted strength. The Nanai were river people, fish people. Like the taimen, they were endangered, and included in *The Red Book of Peoples of the Russian Empire,* with only fifteen thousand or so of them remaining. They were in this area long before the Russians and were still the true guides. The Nanai, Oleg said, knew where the fish were.

Ivan steered the raft standing up, leaning into the oars with his whole body. He concentrated fully on the water, shifting course to navigate logjams and tugging currents as we swept around the huge bends in the river. I noticed that the forests along the riverbank were filled with water. After a while, we pulled to the shore to let Guido fish. Ivan and I sat on a log above some bear tracks, and he gave me hot tea from a thermos and a handful of sunflower seeds. He gestured to Guido, teaching him without words how to fish the river. The wind gusted and autumn's last leaves twirled to the ground. Guido fished one pool and then another without a bite. We got back into the boat and drifted in silence as Ivan pushed his weight against the river, communicating with small gestures that Guido and I were starting to understand. Here was good water, here was an underwater tangle of logs, here were bear tracks, here were ducks. Following Ivan's instructions, Guido caught two smallish taimen, but they were still big, beautiful fish, long and light green with shimmering, iridescent sides.

Guido was heartened, but not buoyant. These two baby taimen were signs that he was on the right track. But it was like finding pottery shards above a pharaoh's tomb. The mother lode lay somewhere below him. He was standing on top of it, but he didn't know exactly where it was. He was beginning to wonder if these taimen operated on some mysterious clock, switching on and switching off. They were either biting or they weren't. What triggered them to bite, he didn't know.

Ivan was surprised by Guido's baby taimen, like an adult watching a toddler who had learned to build with blocks. After Guido's second fish, I turned around to find him shadowing Guido's fluid cast, as if testing to see how the motion felt in his body.

The rafts met up for lunch: soup with a chicken thrown in. It was hot and delicious. The fishing reports were modest; Guido was the only one who'd had any luck. He underreported his victories and ate lightly. Around the table, the fishermen were preoccupied and quiet.

That afternoon we were beached on a side channel when Guido yelled out. Ivan and I rushed to his side and found him visibly rattled. A monstrous fish had lurched out of the water and come at him like a torpedo, mouth wide open. Ivan stared at the calm water and pantomimed that the taimen must have been following Guido's fly, and when Guido pulled his fly to recast, the fish had lunged after it. We got back in the boat and continued downriver. Guido was haunted by the event, by the shock of it. How long had the fish been following his fly? He revisited the moment again and again, as if to resolve some-

thing irresolvable. "These fish couldn't care less about us," he concluded. "They're totally unafraid." But I could see that the observation
gave him little peace. The fact was, he had no idea what these fish
were like.

At dinner we shared tales of our day, huddling close around the
dining table with our vodka and soup. Guido clinked his glass to
make a toast. He had become Russian in this way, and delivered a
heartfelt and sweeping tribute that spoke to history, our time, the
beauty of our earth, the extraordinary moment we were sharing. He
was in his element, relaxed and happy; there was no division within
him about his role, the things he should be doing or the places he
should be. In this adverse setting, he was at rest. His voice softened,
and throughout dinner he told tales of his uncertainty and missteps,
of his doubts about landing either Abramov or a trophy taimen. He
was both witty and guileless, quick as a trout, teasing himself one
moment, baring his soul the next. Happiness ran through him like a
current; we could all sense it. And, for an hour or more in the cold
night, we found ourselves rising to meet him, tilting our wings at his
higher altitude. In the frozen middle of nowhere, we too felt a giddy
sense of freedom. This, I realized, was where Guido won his biggest
battles.

I managed to temporarily forget my worries, but the next morning
I was faced with the strange reality that the tools of my minimalist
trade were being taken from me, one by one. I couldn't write because
of my hand, and now it seemed our electronic devices wouldn't charge
on the little portable generator. With a dead phone, I couldn't record
conversations or take photographs. There was nothing to do but remember things as best I could. The fact that I was physically vulnerable was changing my relationship with Guido. I had never needed
him like I needed him now. Yet of all the men on the trip, I now felt
the least safe with him, for he was wholly focused on taimen. He had
not asked about my hand. It was possible he hadn't noticed that I
couldn't open a bottle of water, or cleat the anchor. From our ranks,
I had recruited a sleeping-bag stuffer and a shoelace tyer, but my help-

ers were already growing weary of their responsibilities. These men were here to fish; they had left their womenfolk at home for a reason. I was only occasionally a pleasant reminder of domestic life. I was alone and in need and the situation made me both frustrated and unhappy.

That morning I struggled with my frozen bootlaces, close to tears with pain. The only person I could legitimately demand help from was Guido, but I had not wanted to disturb his preoccupation with taimen. I was at once overcome by the absurdity of it all, both his position and mine. I left my bootlaces untied and stomped outside to find my cousin. He was busy gearing up, and I could see his impatience with my request for a moment of his time. What did I want? I knew as soon as I delivered my short, vehement sentences that I had never spoken to him this way. For the first time, I understood what it was like to be Lee.

"I am injured," I told him, holding out my hand. "I need help. I can't ask these men for their help anymore. I am your responsibility. You need to take care of me." I was flushed with emotion. I had never uttered such words to anyone, ever.

Guido stared at my hand, which was swollen and purplish blue. "Jesus," he said. I could see that he felt bad. He looked a little like a whipped dog. "Okay," he said, and he was gone. A little while later he found me before we left. "Thank you for talking to me. That is exactly how you need to be with me," he said. Guido absorbed my message deeply enough to ask after me a few times every day. It was enough to keep me from feeling totally alone.

That day Ivan taught us how to read the river. There were places that had deep holes next to fast, shallow water. This was where the taimen lay in wait for prey. Floating along, we all looked for these places and pointed when we saw them. "You're starting to read the water," Guido told me encouragingly. Ivan's wisdom ran deep. There were patches that looked auspicious, but he responded with a shake of the head. His body rocked gently with the river as he weighted and unweighted his legs, holding his balance while navigating logjams, paddling the oars one after another in a continuous cycle. Once in a while, he exerted all his force to maneuver the raft against the current to reach side channels and secret spots only he knew about.

I got out of the boat whenever we pulled ashore, and headed off to look for things underfoot. I found tracks in the dirt; little paw prints that looked like they might belong to a fox, and the delicate impressions of bird talons, barely etched on the surface. A little farther down the gravel bar, I found bear tracks. They were so huge and deep that my heart started to pound. Nearby was a mound of fresh bear dung, full of red berries. I returned to the paws, which were enormous. I placed my wading boots beside them and saw where they led into the woods. When I looked back to the river, I could see neither Guido nor Ivan. A stab of fear sent me walking quickly back to find them.

As we floated down the river I noticed my uneven perception of time, how the hours slowed and then sped up. With nothing to distract me, I became aware of my cold, hunger, and anxiety. I noted an occasional sense of peace, and long moments of utter silence, when the frigid air didn't offer as much as a birdsong. I saw how these states came and went, as fluid as the river, which I watched closely. Big sweeping gravel bars appeared regularly, and while there was a pattern, I was only just starting to understand how it was connected to the river. It was in the way the water took its direction; how the river and the land pushed and gave way and shaped each other, the flow of the water shifting with each bend. The trees told the story too, clinging to the banks until the force of the water toppled them and they floated downriver, leaving an empty space where gravel from the river bottom was deposited and would build up over time. Soon there would be trees here again, and the process would start over. There were forests and groves of all ages along the banks.

We stared silently as we passed logjams, one after another, like a giant's game of pick-up-sticks. Some went for two hundred yards. Ivan watched them from the corner of his eye, throwing his weight into steering away from them. Sometimes, when we were passing one of these death traps, Guido and I caught eyes. We didn't maintain our connection for long. While we were together in this boat, we were still very much alone.

The next day Guido and I fished with Misha. His boat was small and unstable and a tight fit for three. We were tippy, and we constantly balanced our weight to remain stable. Suddenly it registered that we were without life jackets. How had I not noticed this before? The boat's motor allowed us to explore far into the interior, but fuel was limited so we had to be careful. Misha had become withdrawn playing the guide. In the evenings he had not joined us for vodka toasts but sat beneath our single lightbulb reading on his tablet.

"C'mon Misha," Guido said now to his old friend, "let's fish."

Misha's face came to life, and he bounded into the river as if released from a cage, striding upstream before we were out of the boat. I followed him and stood nearby, watching him cast from a snow-covered gravel bank. He was a small man, but he threw his line as far as anyone I'd seen. It hurtled across the water as if shot from a bow and dropped with perfect accuracy, again and again. Once there was a brief tug, but soon he reeled in his fly. Unlike Guido, whose philosophy was to fish every section thoroughly, Misha wanted to keep moving.

We walked downriver and around a bend to where Guido was fishing, Misha towing the boat as we sloshed through the shallow water. Snow lined the riverbanks, and the trees were almost bare of leaves. Guido pointed for Misha to take the upstream water and I watched them cast together. They were like instruments playing side by side. The wind had kicked up, lowering the temperature what felt like another ten degrees. Even Guido admitted to being cold. We got back into the boat and drifted to conserve fuel. It might have been an hour later that Misha decided to throw the anchor on an offshore gravel bed. He had spied a patch of slow water near the bank. "I like the looks of it," Guido agreed. In the periphery on the river side, a movement caught our attention. A massive tree was floating by, its roots springing up from the trunk like Medusa's coils. We could see only part of the tree's bulk; most of it lay underwater.

"Do you have a knife?" Misha asked me. "You will need one." He handed me a medium-sized knife, well used and possibly handmade. "If we get hit by one of those," he indicated the tree, "you will need to cut the rope free."

I pondered the physics of this as the men headed in opposite directions away from me. I think what he meant was that I could die if a tree snagged our raft. If I didn't cut the rope, I would be dragged downstream into a logjam, where I would be sucked underwater.

The sun came out and glinted brightly on the water. The snow had melted, and when the wind stopped, the day was almost warm. I was feeling dangerously, hypothermically sleepy. I sat up, fighting the desire to close my eyes. A yell from Misha roused me. He was pointing downriver to a bear that was swimming from the far bank toward us. Its head was enormous. I watched its rapid progress across the current. When it reached our side of the river, it hauled its hulking body out of the water and lumbered into the woods without even looking at us. When we left our spot a few minutes later, I looked for it back in the woods, but there was nothing.

We had fallen behind the other rafts. Misha ignored the radio check-ins, the other guides asking for our whereabouts so we could coordinate for lunch. We had enough food for lunch, he told us. I was sad to be missing the hot soup, but I had fallen into a comfortable torpor in which nothing much was important. My head itched. I was five days unwashed. In my relaxed state I imagined that I didn't need the things I thought I needed: heat, warm showers, clean hair. These things had not always existed. There were many people on earth who lived without them. I was, however, concerned about my hand, which was now bruising deeper near the thumb. I turned to the churning, roiling Tugur and these worries receded. I understood now what a wild river looked like. I could see how it behaved. It was free to flow wherever it wanted, yet it adhered to some natural law of symmetry and compensation, like a sidewinder in the sand.

Guido and Misha were comfortably taciturn. They were watching the immense and ever-changing Tugur with tense readiness. To the right, another large fallen tree floated swiftly along, half of its deadly mass submerged. How little it would take, just an underwater snag, to grab the little raft and pull it along.

The sun disappeared and the wind picked up. Suddenly I was wide awake and freezing. I had six layers on top and four on the bottom, and I was still cold. I un-hunched my shoulders and rolled my head around. I arched my back and stretched my sides. A deep breath came spontaneously and I felt better. I told myself to remember to breathe. There was nothing I could do about the cold, or my hand, but I could remember to breathe. I sat on a stump and stared abjectly at Guido, who was casting beautifully to the far side of the current, bare-handed and oblivious to the cold.

The next day, we were with Ivan again. He had become my friend, giving me hot tea and sunflower seeds to crack between my teeth to distract me from the cold. Understanding I was hurt, he wordlessly helped me reassemble my many layers and fastenings after I relieved myself. I grew accustomed to standing in front of him like a child, my waders hanging around my waist as he hoisted my woolens high, hitched my wader straps over my shoulders, clicked my belt and cinched it tight, and zipped up my two top layers. Very little was lost on Ivan, and in this harsh and indifferent place I felt safe with him.

It was in the late afternoon when the river split into smaller strands. Ivan, with a backward glance at the raft behind us, steered hard for the farthest left braid, guiding us through a narrow inlet between two logjams. The water rushed under us, tipping the upriver side of the raft. Ivan planted his legs firmly and used his weight to steady and steer through the current. Suddenly we were in calm water. The Tugur vanished from sight as we pulled onto a sandy bank that divided a slow-moving channel. We got out and Ivan tipped some more sunflower seeds into my hand as Guido set out to fish a patch of eerily quiet water from the bank. He had not proclaimed, as he had before, that it looked like beautiful water. We had not spoken much since lunch, instead watching the massive logjams pass as the river turned this way and that, divided into channels, parted around islands, and rejoined. Now the light had gone pale and it was getting colder again. The feeling of the day had changed.

As Guido took his place in the river, Ivan and I wandered off, Ivan to sit on a fallen tree and I to inspect the shallows upstream. Across the river a Steller's sea eagle circled above a stand of tall pines. I watched it for a while. Apart from crows, we hadn't seen many birds. The migrations were over, with only a few stray ducks and geese passing overhead.

I was not far from Guido when he yelled out. I heard something I didn't often hear in his voice, and on an animal level, I knew what it was. He had a big fish on. I ran to him through the shallows and over the gravel bank, Ivan on my heels. Guido stood like a statue, rod bent like an apostrophe and frozen in silent concentration. He glanced at us quickly. "It's big," he said. We looked at the line where nothing much was happening; the fish wasn't running downriver or leaping into the air. It was as if Guido had hooked a boulder. Then the fish moved, running up and down the opposite bank a few times. Then it stopped. When Guido started trying to reel it in, Ivan made a calming gesture with his hands. Guido ignored him, grappling with his own strategy. "It's really big," he said. "Get the boat, Ivan. We may need to chase it."

I caught sight of Oleg's raft resting on the main channel, across a few inlets and gravel banks. From afar, Oleg could see that Guido had a real fish on, and he came at a run and joined Ivan to instruct Guido on how to bring the taimen in. Oleg spoke quiet instructions while Ivan kept silent. They moved in perfect unison as they stalked the water behind Guido, articulating instruction with their hands. Calm, calm, they gestured, just keep backing up into the channel. This was a critical step, the moment when anything could happen. The taimen was coming in, this was good. But if it decided to fight, there were many ways it could win. It could wrap the line around a log, it could dive deep, it could run. If it ran, it could break the line and even the rod. It could also simply shake its huge head with such violence that it dislodged the fly or broke the leader. Some taimen fought only when they got to the bank.

These were things Guido had heard but did not know firsthand. A power struggle ensued between him and the guides. Guido wanted to do things his way. Oleg and Ivan communicated with head shakes that his ignorance could lose him the fish. They instructed him to back up and drag the fish farther into the quiet water in the left channel. His best chance of landing the fish was to lead it calmly to shore. Guido was no longer listening to the guides and seemed unaware of them; he had a different scenario in mind. He wanted to bring the fish into a small inlet on the right side of the gravel bank, closer to the main channel. Oleg and Ivan responded that this was a very bad strategy. Guido's instincts were not serving him here, with this fish he didn't know. But it was his fish, and the conversation between them belonged to no one else. Everything in Guido was attuned to the taimen's movements, the little refusals and surrenders on the end of the line as it came haltingly closer. Maybe he was remembering Abramov's most emphatic advice: do nothing to anger the fish, do nothing to hurt it. If it did not register pain or hostility, the fish might come to you.

For Guido it was a disconnect. He knew how to deal with a fighting fish. This fish continued to come in slow and heavy, locking Guido into a kind of trance. There was no guarantee his line would hold, or that there wasn't an underwater log the fish could decide to dive under. There were still many ways this could go wrong. But the fish continued to come in, its size transmitted through Guido's body in the angle of his limbs, the near crouch of his back, in all his muscles working in opposition to its weight.

I saw the conversation change then; it was in the way Guido gingerly backed up, swung his rod to the right, bending it deep, and gently throwing the fish off balance to tire it out. The taimen was close now, and Guido's focus was entire. It was as if he were pleading with the fish: I want to save you, but first I must catch you. Come with me, just come. You will soon be free.

The line and the fish beneath it moved into the dark water of the channel, just ten yards away. Guido had surrendered to the unfamiliar strategy of gently leading the fish deeper into the channel. He kept his rod bent at the same angle, dragging the taimen to the shallow water, where it would be forced to surface. I could tell it was strange for him, this slow capture instead of a battle royal. We watched the water as Guido's line moved and whatever was at the end moved incrementally with it. We could see nothing in the dark water below. We waited as he pulled the taimen farther and farther in, no one speaking as we watched to see what would emerge from the water.

Finally, the smooth surface rippled and the water was broken by a fin that was followed by a broad back. "Jesus," Guido said under his breath. We stood staring at the amount of water that was displaced. The others exclaimed softly, not wanting to break Guido's concentration. It was a sacred moment. Oleg and Ivan were knee-deep in water, following the fish from behind. All at once, the fish surfaced, and in the same moment Oleg and Ivan had their hands under it in the shallow water.

It was a very big fish. There was shouting all around. This was it, the world record. Everyone knew it. The guides were both shocked and delighted at the unlikely victory. It was a trophy taimen caught on a bit of fluff and feathers. Guido led the fish farther into the shallows, where it could be controlled. After measuring and photographing it, they eased the fish into a specially made porous sling so it could be weighed. The taimen looked strong and extremely well fed; even with two men, it was a strain to lift. Its back was a beautiful olive green, with sides gleaming with a rainbow of colors. The scale registered the fish at close to seventy pounds. I wondered if this was the first time the taimen had encountered a human being, and how its life would be different now. After it was released, a new memory could impact its choices. This fish might stalk humans next. It might never be caught again.

After it was properly photographed, Guido led the taimen gently back into the current and pointed its head upstream. We watched as it opened and closed its mouth, taking in the oxygenated water. Guido was barely holding it when the taimen regained its strength and swam slowly back into the depths. He stared after it, unsettled for the second time that day. It was disconcerting that this monster fish had been relatively easy to land. It hadn't been so much a fight as an offering.

That night there were toasts and the Russians built a hunter's fire. Of the three kinds of fires, this was the biggest, constructed of large branches and small trees laid atop one another, end to end. It was an honoring, and it burned like an inferno in the inky, frigid night. Standing near it, our bodies started to release the tension of the cold. The Russians waved us back from the flames; our socks and shoes would melt. We moved back only slightly. Transformed by the heat, it was hard for us to care about such things. The warmth it gave us lasted until we were in our tents, where we would face the coldest night yet. By morning everything in camp was covered with frost and every liquid was frozen solid.

Our last camp was the most beautiful, built on a broad gravel spit where the river opened into a small delta. There were countless little channels braiding through the sand, and the view was wide and magnificent. There were snow-covered mountains to the east, pink with the setting sun. Guido had lost his last fly to the river, and he sat at our folding dining table with his vise and bags of feathers and fur, tying more.

It was even colder that night, and crystal clear. Again, we drank vodka for the feeling of warmth. Guido toasted Misha and their more than twenty-year friendship. The talk after dinner was of death, of fatal accidents these men had witnessed on rivers. Stories were told of lost wives, children, friends. A father and son had died together after being told to wedge themselves into the bottom of their raft before going over a rapid. When the raft flipped, they got tangled in their fishing lines and couldn't swim to shore. A newly married man followed his wife into a logjam and both drowned. Even the Tugur had lost someone to a logjam. These watery deaths were frightening to me. It was not at all how I wanted to die. It was then that Guido admitted that the Tugur was the most dangerous river he'd ever floated. I stared at him, realizing he'd withheld this observation until now, for my benefit.

"Why?" I asked.

"There is no margin of error," he said. "If you're off the boat above a logjam, that's it." Through my mind I replayed the moments when we nearly tipped over, of how close we had been to death. Guido turned to Misha. "And we don't have life jackets."

"There would be no point," Misha said.

That night we all drank too much vodka. I knew how dangerous this was, but I could not resist the illusion of warmth the alcohol was giving me. A few of us stayed up talking by a small fire near the river. As the temperature dropped, I felt relaxed and content, and I stayed there until the fire no longer warmed me. That night, in the subzero temperatures, I was colder than I'd ever been. My sleeping bag was cinched tight, with only a small hole for my nose, which was numb in the frigid air. In the end, I was too cold to sleep. I lay awake, shivering, the icy temperature sinking into the marrow of my bones. At one point I felt something else creeping into me, through a crack in my defenses, like a virus or a flu, but deeper and darker. I wondered if death felt something like this.

I choppered out alone while Guido and the others floated on. It was always the plan for me to leave a few days early, and I was grateful. The ride back to Khabarovsk took seventeen hours in a little car driven by a twenty-year-old, Andrei, who eyed me with worry in the rearview mirror. My hand was swollen and a bright red line marched steadily up my forearm. The warmth of the car had unleashed the pain and inflammation. "Is this red alert?" Andrei asked. I wasn't sure what difference it made, as we had no reception and could go no faster. I slipped into fever as we jounced along. Every few hours Andrei pulled over and gave me a tablet of something, which helped me sink below consciousness. At some point I was aware that he was refilling his gas tank with a large plastic container from the trunk. The sun was setting, turning white mountains to rose, and in the twilight Andrei told me how much he loved this land.

Close to midnight, we passed through a mountain range and Andrei woke me because the air was filled with butterflies. He had never seen this before. I saw them too, delicate and flickering white in the darkness. He said he would call them "ice butterflies" because there was often black ice on this pass. He talked to me as he drove, not wanting to be alone with the ice and the butterflies. I tried to stay awake with him, watching the butterflies flitting in and out of our headlights. A little while later, we saw a red fox with a thick bushy tail and we slowed as it trotted toward the trees. Before it disappeared into the forest, it paused and looked back at us boldly, as if it knew we posed no danger. We drove on, agreeing that it was a beautiful fox.

We reached my hotel at four A.M. The receptionist called for a doctor, and soon I was visited by a man in green scrubs carrying a metal box that looked like a tool chest. As he set it down, I could hear glass vials clinking inside. After inspecting me, he wrapped my hand, instructed me to bare my bum, and stabbed me with two big injections from his glass vials. I imagine they were penicillin and some kind of painkiller. I was past caring. I slept for a solid twelve hours, and when I awoke, my hand had lost its angry redness. I was finally warm. Out my window was the great Amur River, which forms the border between Russia and China, and in the bare trees a flock of tiny birds flitted in the branches. I wished I could speak with my mother, who had traveled the world looking at birds and might hazard a guess at what brave—or foolhardy—little birds migrated so late.

I was back more than a week before I knew how to write about the Tugur. But I dreamed about it. Uncomfortable dreams with anxiety and occasional majesty. I would continue to think about these ancient fish, and how long they'd been on earth, and how little was known about them. Ivan had told me that taimen had been seen feeding in the estuary of the Tugur, near the mouth of the river. Did they also venture to sea? Months later, when I asked Alexander Abramov about it, he shrugged and said there was no mystery. It was simple; it was about food. Maybe some years when the chum were lean, the taimen went to the sea to eat mollusks. To a Russian scientist, there were few great mysteries in life, and whether taimen went to the sea wasn't one of them. It was to me. Taimen reminded me of another fish that had the ability to shift its behavior, its very physiology, depending on the conditions. Taimen reminded me of steelhead.

That week the Tugur began its future as a world-class fly-fishing river. Within hours, Guido's record was broken by Ilya Sherbovich, who had also come to Abramov's camp to try his luck catching a trophy taimen on a fly. Mark and the others caught big fish in their final days, though none as big as Guido's. Some basic code had been cracked, and Abramov was pleased. He invited Guido to return the next year, when Guido caught an even bigger taimen. That week, in the calm water of a side channel, he saw a fish that was well over a hundred pounds pass beneath his boat as it chased a salmon. Guido had never seen a bigger freshwater fish. Given the chance, he told Abramov, he thought he could land one of these giants, though Ilya might get there first. That the Tugur might become one of the most challenging, dangerous, and exciting fly-fishing rivers in the world meant one thing above all else to Guido: the river would find protection.

CHAPTER 30

GOING FORWARD

L ATER THAT YEAR, Alexander Abramov was in New York City for business, and Guido invited him to the exclusive Knicker-bocker Club. The club had a dress code. Guido was worried that Abramov didn't know or, if he knew, might ignore it. Guido had made sure to wear a suit and arrived right on time. Abramov arrived a little late, dressed casually in jeans. He sat down, kicked his feet up on a table, and grinned.

"Guida, you're all fancy in a suit."

Guido grinned back. "I'm trying to impress you!"

They talked warmly about the coming season. Abramov had invited Guido back to the Tugur, and wanted him to outfit his lodge with fly-fishing gear. Abramov had also agreed to invest in a number of radio transmitters to investigate the only mystery about taimen that interested him—how taimen were affected by being caught. Did the fish simply swim away and resume life as usual, or were they somehow damaged or altered by the experience? He had heard different myths—taimen were strong, taimen were delicate. He didn't know which, if either, was true. Abramov's biggest concern was that some of the fish died after being caught. If catching them was some-how traumatic, he wanted to know. Perhaps he also suspected that fly-fishing was easier on the fish.

Guido pressed Abramov to consider supporting an effort to protect the neighboring watershed, a proposal that Abramov instantly re-jected. It was a watershed with many subsistence-level poachers, peo-ple just trying to survive. The government would never apprehend them, nor should they. Russia was a hard place to live, hard for fish and for people. The Tugur was another matter. Abramov said, "I want you to fish this river every year for the rest of your life."

It was a victory beyond the price of rubies, and one that Guido would take.

Alexander Abramov had become a founding member of the Rus-sian Salmon Partnership along with Ilya Sherbovich and Vladimir Rybalchenko. The organization had committed to education pro-grams in Russia that promoted protected habitats and modern fish management. Russian children would be taught about salmon, and how to fly-fish; they would be taught the curious practice of catch-and-release. There was a chance that the new generation might see salmon and rivers differently and fight to protect them.

And Guido would be able to return to the Tugur, this time with his son, Gee, who would finally see Russia and the rivers his father loved so dearly.

The bridge to Alexander Abramov had been built just in time. Early in 2015, President Putin signed a bill that allowed the federal govern-ment to target and prosecute "undesirable" NGOs working in Rus-sia. Those employed by such organizations could be sentenced to as much as six years in prison. Passed by both houses of the Russian parliament, the law was intended to protect the Russian state from "destructive organizations" that threatened the constitutional order of the Russian Federation. There was an international outcry against the law, which Amnesty International declared threatened "funda-mental freedoms." Soon thereafter, one of the Wild Salmon Center's primary partners, Sakhalin Environmental Watch, was declared a foreign agent for accepting support from the Wild Salmon Center as well as environmentalist actor Leonardo DiCaprio.

The remaining Western NGOs still active in Russia began packing their bags. The foundations followed suit, one after another—the C. S. Mott, MacArthur, Walton, and Gordon and Betty Moore foundations—all of which had a host of legitimate reasons to turn their attentions elsewhere. The Wild Salmon Center was still very much alive in Russia, but it was increasingly alone.

In 2015, the Moore Foundation had its wrap-up meeting with the Wild Salmon Center and the other conservation groups that had been part of its decade-long salmon initiative. There had been a bitter battle within the Moore family and within the foundation about whether to continue supporting the program. Gordon and his children were openly proud of the mission; as fly fishermen, they loved wild rivers and recognized the value of protecting salmon strongholds. Ultimately, though, they decided to move on to other beleaguered species: shrimp in Thailand and tuna in the South Pacific. Guido protested this decision for two reasons: these were not keystone species, and they were far from home. Shouldn't they protect their own backyard first? The consensus at the Moore Foundation was that the future in Russia was too uncertain.

There was a new program officer at Moore for the salmon initiative, and she led the meeting as grantees from different regions of the Pacific Rim enumerated their accomplishments. It was a dry accounting presented by people who didn't remember the origins of the program and who had never seen the rivers they were talking about. Guido listened restlessly, wishing they could bring the rivers to life so the audience could see and feel them. The heart and soul were missing, and the Wild Salmon Center's hard-won achievements in the Pacific Rim sounded like administrative details. As the meeting went on, Guido grew frustrated. Every one of these details was a story, a story of a wild river, of brave people, of discovery, trust, and partnership. He listened as long as he could and, ignoring protocol, jumped in at the end of the final presentation. It was his last chance to make a plea, to make these people with spreadsheets in front of them see what was truly at stake. The conservation leaders the Wild Salmon Center had coaxed out across the Pacific Rim were finally hitting their stride. The Moore salmon initiative had nurtured and connected them, and helped create a movement. Without foundation resources to support them, these groups would be left isolated and vulnerable.

"Fifteen years ago there were all these lonely little groups doing battle out there," Guido continued. "We were able to connect them, breathe life into them. Doing this unleashed a new level of conservation. We've done so much—and now that we've built real momentum and reached this critical moment, everyone's leaving. It's the foundations that need to have the courage to stay in tough places and remain engaged; they need to lead.

"I know there's a risk," he continued, "but we have to keep these networks alive. The political winds keep changing, but the rivers are the same. We need to transcend the political—to have conservation be the shared language between our countries. This is a beginning; this is something we can agree on. We must seek out Russians to fill the support gap. Finding funders like Alexander Abramov might not work out, I realize. There is no crystal ball. But to walk away because we can't predict the future would be disastrous. We cannot leave this thing we've built. We cannot completely abandon the communities we've forged."

Guido called me later that day. He was down. The natural world didn't work in ten- or fifteen-year increments. If only these people could see what he had seen, if only salmon and rivers could speak a language they could understand.

There were those who recognized the significance of what the Wild Salmon Center was trying to accomplish. The Yale School of Forestry and Environmental Studies, Guido's alma mater, invited him to speak about strongholds multiple times, always to a packed auditorium. Stanford business school made a case study of the Wild Salmon Center, linking its remarkable success to its adaptability and unique willingness to consider any and all available solutions—not unlike the fish it was fighting to protect. When speaking at Stanford, Guido watched his audience grow wide-eyed at his tales of how things got done in the Russian Far East. He laughed, knowing his was a business model that broke the mold.

There was no question that the achievements of the Wild Salmon Center and its local partners were extraordinary. Across the Pacific Rim, millions of acres and tens of thousands of river miles of habitat had been moved into protected status. The Wild Salmon Center had formed new organizations, built coalitions, convened international symposia, authored dozens of scientific papers, and brought conservation groups from Russia, Canada, Japan, and the United States together for the first time. With its partners in Alaska, British Columbia, and the Pacific Northwest, the Wild Salmon Center made a series of conservation wins, including new laws and policies restricting logging, mining, and hatchery fish production along thousands of miles of streamside habitat in Oregon and Washington State, the defeat of a major liquefied-natural-gas terminus on the Skeena River in British Columbia, and the defeat of a hydroelectric dam in the Susitna River in Alaska. More than 52 percent of all Pacific salmon fisheries were now certified as sustainable by the Marine Stewardship Council, or in a Fishery Improvement Project to obtain certification, including most of the major salmon fisheries on the Kamchatka Peninsula. The commercial fishermen of Sakhalin Island had created a thirty-river, 1.8-million-acre wild-fish zone. With its Russian partners, the Wild Salmon Center had won the designation of six new national or regional parks in salmon strongholds totaling 2.7 million acres, and another 4 million acres encompassing four pristine salmon rivers were proposed for protection. In 2017, Alexander Abramov deepened his commitment to the Tugur by purchasing the logging rights to prevent any future clear-cutting and road building.

The one great loss was Bristol Bay. Outspent by the opposition, in 2018 the Wild Salmon Center and its Alaska partners lost a ballot initiative campaign to rewrite salmon river conservation rules, which would have protected the region. The only hope now was to fight the federal and state permits to delay drilling until the next presidential administration.

Guido's singular achievement was in raising the issue of conservation above geopolitics. Even as diplomatic relations continued to deteriorate, Russia and the United States strengthened their commitment to protecting salmon and the pristine habitat of the Pacific Rim. On October 5, 2015, Russian prime minister Dmitri Medvedev signed a new Russian Federation law allowing the creation of federal fishery protected zones on important rivers in Russia. The Wild Salmon Center's Russian colleagues had been pushing for this for almost a decade. In fishery protected zones, practices such as logging, mining, and road building could now be banned. Then, in 2016, top leaders from US and Russian wildlife agencies agreed to continue cooperative work on a broad array of conservation and wildlife science initiatives, as part of a renewed pact. Work would include polar bear research and conservation, marine mammal studies, and migratory bird monitoring and protection. The agreement endorsed a series of ongoing science and technical-support conservation projects between the Wild Salmon Center and its Russian partners.

In terms of Western support, however, Guido and his staff were on their own again. But alone was old ground; alone was what they did best. By now, Guido had hired people who were at the top of their field, and he relied on them completely. He stepped back and took some time to think about what to do next. He had, as always, been considering the situation from above. What he saw in the numbers alarmed him. According to projections, in the next thirty years the global population would reach nine billion people—full capacity. After 2050, this number was expected to taper off in response to the earth's limited resources. As they had seen with salmon in the Pacific, there were only so many trees, only so much water—only so much food. Commensurately, in that time, the world's demand for seafood would increase by 70 percent.

Guido was focused on a critical aspect of this projection. In fifteen years, the emerging middle class was expected to double—Asia's population in particular, and it would be demanding quality sources of protein. The Pacific Rim held salmon rivers that could provide a large and steady source of protein, but the fish would have to survive the approaching gauntlet of increased consumption. Over the past twenty years, the Wild Salmon Center had learned how to protect rivers, but to extend that protection until 2035 required resources they didn't have. Guido and his board needed to raise enough money both to support groups on the ground and to fend off threats that might arise before 2035. He needed to create a war chest. If Guido and his staff could raise an additional $30 million, they could focus on the fight and not the funding. They could effect legislation and add new parks and streamside buffers while they strengthened local groups. Guido believed it was the only strategy that could protect strongholds not just for the next generation but until the end of the century. Once he had settled on a plan, he was focused, calm, and utterly determined. The Moore Foundation came around and started him off with some funding and a commitment to set aside resources for future grants in Kamchatka. It was a start. Soon Guido was on the road again, taking remarkable people to remarkable rivers and letting the dots connect themselves.

That spring we were on our way to the Deschutes, and Guido pulled off at the Bonneville Dam. He wanted to show me something. We were towing his little skiff up the Columbia Gorge, following the mighty river into which so many other rivers flowed, including the Deschutes. I thought about how when Lewis and Clark navigated the Columbia in 1805, it was teeming with salmon. Now only a fraction of their original numbers swim upriver every year, navigating the enormous Bonneville Dam by fish ladder.

Guido took me to see the viewing windows, where the fish passed through the top of the ladder behind oversized glass windows. We stood and watched while hundreds of salmon of all kinds fought their way against the current like so many stalwart soldiers marching home. They were beautiful and strong, and for some reason their urgent, silent efforts brought tears to my eyes.

The senior engineer who built the dam for the US Army Corps of Engineers in 1934 initially refused to make any allowances for the salmon, declaring he did not intend to "play nursemaid to the fish." The ladder was installed later in the forties, and immediately the fish, thwarted for years by the impenetrable barrier, came streaming in. As new legislation required the removal of dams from rivers up and down the Pacific Northwest, the story was the same. The fish came surging back. It was as if they had been there all along, waiting.

We didn't stay at the ladders for long, because Guido had an agenda. He wanted to show me a secret place. He had overcome a powerful inner resistance to protect this place from even one more person knowing about it; it was that precious. I swore not to disclose any particulars about its location. It was, as he put it, one of the last places on the Columbia where the river was still like it once was, and he wanted it to stay that way for as long as possible. Also, he didn't want the competition finding out about it.

It was a vivid, sunny afternoon, and we motored up the Columbia against a warm wind. The sheer amount of water flowing by was staggering, and despite the obstructions of man, nature abounded. We jounced rhythmically against oncoming waves as ospreys and bald eagles circled overhead, their shadows crossing our boat. Cormorants scooted downriver, their quick black wings flapping just inches from the water. We occasionally passed tackle fishermen with their boats anchored in the deep current.

A mile or so on, we veered off the main channel to a wide expanse of water that flowed north into the interior. Here, the Columbia transformed. Slowed by small islands, it meandered over pools and scuttled over cheerful riffles. Guido cut the motor, and we drifted in the calm water. The wind filled the silence as it rustled the tall grasses and the leaves of cottonwoods and alders. A family of mergansers paddled past, and the occasional blue heron stalked in the shallows. For a while we said nothing.

Then Guido took up the oars and rowed to his spot. It was time to get to work. He instructed me to stand in the bow of the boat, where I was to stare into the river and look for fish. "They're hard to see," he added.

It was more than hard. The bottom of the river was barely visible; medium-sized rocks came in and out of focus. In addition, we were drifting at a different pace than the river was flowing, while the bottom wasn't moving at all. The wind rippled the surface of the water, creating glints and shadows. My eyes were confused, caught by so many layers of movement and light. What I was meant to be looking for were "springers," the first, strongest, and most sought-after chinook of the season.

"They won't look like fish," Guido said unhelpfully; "they're more like a mirage." The instruction continued. "Look as far out as you can," he said. "Focus and then soften your gaze. You'll see them in the periphery, like shadows, like a suggestion." We drifted in silence. I saw nothing. I was a good spotter, but this challenge was beyond me. Fifteen minutes passed with more nothing. "Your brain doesn't know what to look for yet," he said. "It's like learning a new language. It takes time." Nothing and more nothing. "You just need to see one," he said. "Then you'll understand the pattern of the shape and movement." My brain felt slightly addled. His instructions weren't making total sense, but I knew he was trying to describe something new, a place I had never been, and while he could point me in the right direction, I had to make my way there on my own.

Then, next to me, he quietly announced, "There's one." Then, "There's another." I followed his gaze and saw green water and rocks. "There, and there. Oh wow, did you see the blue of his back?"

I saw nothing, but I was completely and happily absorbed in trying. The time passed. We decided to swim and then nap on a nearby island while the sun moved to a better angle. As Guido drifted off to sleep he told me that Lee had called him in Russia recently and reported how Gee had taken off one morning at the Deschutes before dawn. By fifteen, the wilderness had taken hold of his soul, and his love for wild places was steadily growing. Gee was an excellent fly fisherman. He was also a hunter, like his grandfather, who had loved to hunt chamois in the Alps. It was deer season and the boy had his sights set on a mule-deer buck. One could try for days and weeks to find a worthy buck, roaming the hills far and wide. It wasn't until you learned how to become part of the land itself that you could even get close to such a creature. The chances of a kill were small indeed.

Lee didn't panic when Gee didn't come home. The times when she felt the need to panic had long since passed with her boys. It wasn't until late in the afternoon that he returned. His face was aglow and he seemed transcendently happy.

"How was it?" Lee asked.

Gee hadn't seen a buck, but that hadn't mattered at all. He saw golden eagles looking for prey. From his perch high in the rimrock, he could look down and watch them hunt. He saw big-horned sheep, so close he could see them feeding on the tall grasses of Mutton Mountain. Down toward Eagle Creek, a cluster of turkey vultures had circled a kill. He saw mule-deer does and their fawns. He watched the sun rising over the canyon, turning the river to gold. Retelling the story, Guido got teary. His son had learned something from him, perhaps the most important thing he had to teach. It wasn't about killing a deer, or catching a fish. It was about a connection to the earth; it was about solitude and wholeness. It was about discovery.

We went back to the river, swishing through the tall grass and the insect hum, to the little boat resting on the sand. Guido rowed out into the current, and I got back in position on the bow, staring down, in the hopes of glimpsing something I hadn't seen before. The sounds of the river soothed me as my eyes began to decipher things that might or might not be suggestions of fish. It was like it used to be with us. While a small part of me was working hard, most of me was at rest. I was perfectly content. I forgot the fish for a moment and looked into the river. Part of the Deschutes was flowing by us. I felt both overwhelmed and slightly hypnotized, as if time were standing still. We were children, now we were parents. All too soon our cycle would come to an end, and another would begin. It was sad, and it wasn't sad. Why should we be any different from every other living creature? How lucky we were to be a part of it, even for a moment. I realized

Guido had always known how to live. More than ever I believed that he had taken his instruction from salmon. To live life to the fullest, swimming fast and hard and dying quickly. To not live a half life, to give everything to the journey.

Guido moved the boat into the current, and the clear water moved fast beneath us. I stared, trying to adjust my eyes, focusing hard on the rocky bottom, which was difficult to see with the reflections of light from above and below. My eyes were torn between wanting to find shapes and following the current. For a moment I surrendered to watching the water. It was lovely, reflecting the afternoon sun that winked off the white rocks resting on the riverbed. I wondered if there were as many white rocks as there were gray, and I wondered what made a rock white or gray. Then, below me, moving across the white and gray rocks, I saw an unmistakable shape advancing against the current, upstream. It was big and looked very much like a fish. "Guido!" I cried.

"What did you see?"

I knew to tell him exact details. "A bit of chrome body and the dark blue of the back. And a black-edged fin." He smiled. "Good," he said, turning back to the river. "Now you have it."

ACKNOWLEDGMENTS

THIS WAS A LONG and complicated book to write, and I am grateful for the many people who contributed to the process, starting with those who agreed to be interviewed, gave so generously of their time, and helped me understand, piece by piece, how to tell this story. There was much I had to learn about fish, rivers, conservation, Russia, and the geopolitics of the Pacific Rim, and I couldn't have had better teachers than Misha Skopets, Xanthippe Augerot, Mariusz Wroblewski, Emily Anderson, and Leila Loder. There were many who chose not to be named in the book for fear it would compromise their work. I want to thank them for the risks they took in opening up to me and for offering me their invaluable contributions.

A special thanks to Spencer Beebe, Laurie Malarkey Rahr, Sarah Rahr Fortna, Tina Rahr Lane, Sam and Susie Bell, Starling Childs, Steve Beissinger, Stephen Kellert, Serge Karpovich, John Sager, Pete Soverel, Jack Stanford, Nick Gayeski, Jeff Mishler, Mike Finley, Mike Sutton, Jim Wolfensohn, Peter Seligmann, Vladimir Burkanov, Aileen Lee and the Gordon and Betty Moore Foundation, Roman Sartrov and Sakhalin Environmental Watch, Maxim Ageev, Nikita Mishin, Ilya Sherbovich, John Kitzhaber, and Alexander Abramov.

I want to thank the employees of the Wild Salmon Center, particularly David Finkel and Hollye Maxwell, for fielding my questions and tolerating my regular presence these past years.

I am extremely grateful to John Childs for showing me Bristol Bay in all its splendor and for supporting this project so generously. Thanks also to the amazing guides and bush pilots of Crystal Creek Lodge, who are top-notch, and whose teachings—on the river and in the air—opened Alaska for me, and made me want to return, again and again.

Thanks to Kris Timkin, for understanding this project and sponsoring a key expedition, and to Bobbie Bristol, for her early support. Thanks also to my patient and intrepid transcriber, Sarah Mason, who labored through more than a hundred hours of interviews, deciphered challenging accents, and persisted through all manner of background noise. Thank you, Deborah Seigel, for your early encouragement and clear voice, and to my dear friend Joanna Goodman, for her warm heart and poetic ear. A loving thanks to Dayna Goldfine and Dan Geller, my trusted sounding boards, and Lynette Walworth, for one morning making me draw it out with pencil and paper. For beloved Alexis Schaffler Thomson and our years down the hall and in the trenches together; you were my comrade through this. For dear Kelsey Hennegen and your pure spirit, careful reads, and steadfast support. Thanks to Mark Becker for his help on the home front, and to Mengfei Chen, for being so calm and efficient, and to Spiegel & Grau and Penguin Random House for their excitement and support of this book.

Much gratitude to Hilary Redmon and Molly Turpin for getting me to the finish line, and to the enduring patience and talent of designers Caroline Cunningham and Greg Mollica. A special thanks to Janet Wygal, for her exacting eye and attention to detail.

I want to acknowledge three departed elders on whom I often called for guidance in the quiet (and sometimes desperate) hours:

Peter Matthiessen, Galway Kinnell, and my father, Tom Malarkey. Your spirits are ever strong, and steadied me in rocky times.

A huge thanks to Ann Banchoff, my greatest cheerleader, for giving consistently brilliant feedback, inspiration, laughter, and steadfast support. You've understood and loved this book from the start, and have urged me on every step of the way. Thank you for shoring me up and making this process less lonely. A warm thanks to Christina Campobasso and Scott Morehouse, for living three doors up, for the many days and nights of friendship, for cheering me on, for feeding and watering me, for always being there.

Thanks to Cousins Will Fortna and Henry Easton Koehler for their keen eyes and ears and all-around editorial smarts—and to Cousin Leiv Fagereng for painting the amazing fish for the galley cover. I am blessed with family, which brings me to my closest people: I am indebted to the brilliant and loving Sarah Malarkey, Nicole Newnham Malarkey, and Thomas Malarkey—for listening, helping me sort through it all, and being so consistently present. I don't know what I would have done without you all. Thanks to my mother, Brent Malarkey, my toughest critic and staunchest supporter—thank you for showing me such extraordinary places, and for laying the foundation for my life's work. Loving thanks to my son, Elliott—for enduring my absences and struggles and buoying me with wisdom and humor. I don't know where to begin in thanking Lee Rahr and her three wonderful circles—Guido Jr., Sumner, and Henrik. You have shown me endless patience, love, and support. This book wouldn't have happened without you.

I am beholden to Eamon Dolan for his early editorial contributions and for helping me understand my own vision. Svetlana Katz, for your tough eye and honest words; because you always pick up the phone, because no question is too small. To Cindy Spiegel, thank you for wanting this book from the get-go, and for taking me on so late; I was beyond happy to have you as an editor and treasure your friendship.

A profound thanks to my human stronghold Tina Bennett, for sitting in my kitchen years ago and listening to me describe a book I barely knew how to write, and believing in it—and me—completely. You have been an unflagging companion, a fellow swimmer as we've made our way upstream. Thank you for trusting me, for changing course, for helping me find my way home.

Finally, I want to thank Guido Rahr, the subject of this book, my guide, cousin, collaborator, and friend; you never wavered during this process. Thank you for your patience, generosity, and spirit—and for continuing to offer such precious instruction.

BIBLIOGRAPHY

I also want to acknowledge the following publications, which were all integral to my research:

Applebaum, Anne. *Gulag: A History*. New York, NY: Anchor Books, 2003.

Arsenyev, Vladimir. *Dersu Uzala*. Honolulu, HI: University Press of the Pacific, 2004.

Augerot, Xanthippe. *Atlas of Pacific Salmon: The First Map-Based Status Assessment of Salmon in the North Pacific*. Berkeley and Los Angeles, CA: University of California Press, 2005.

Ford, Corey. *Where the Sea Breaks Its Back: The Epic Story of Early Naturalist Georg Steller and the Russian Exploration of Alaska*. Boston, MA: Little, Brown and Company, 1966.

Gilbert, Martin. *Atlas of Russian History*. New York, NY: Oxford University Press, 1993.

Kennan, George. *Tent Life in Siberia: An Incredible Account of Siberian Adventure, Travel, and Survival*. Layton, UT: Gibbs M. Smith, 1986.

Lichatowich, Jim. *Salmon, People and Place: A Biologist's Search for Salmon Recovery*. Corvallis, OR: Oregon State University Press, 2013.

———. *Salmon Without Rivers: A History of the Pacific Salmon Crisis*. Washington, D.C.: Island Press, 1999.

Mallaby, Sebastian. *The World's Bank: A Story of Failed States, Financial Crises, and the Wealth and Poverty of Nations*. New York, NY: Penguin Group, 2004.

Meehan, William R. *Influences of Forest and Rangeland Management on Salmonid Fishes and Their Habitats*. Bethesda, MD: American Fisheries Society Special, 1991.

Montgomery, David R. *King of Fish: The Thousand-Year Run of Salmon*. Boulder, CO: Westview Press, 2003.

Prud'Homme, Alex. *The Ripple Effect: The Fate of Freshwater in the Twenty-First Century*. New York, NY: Scribner, 2011.

Safina, Carl. *Song for the Blue Ocean: Encounters Along the World's Coasts and Beneath the Seas*. New York, NY: Henry Holt and Company, 1997.

Sager, John. *Uncovered: My Half-Century with the CIA*. Bloomington, IN: West-Bow Press, 2014.

Sakurai, Atsushi. *Salmon*. New York, NY: Alfred A. Knopf, 1984.

Shpilenok, Igor. *Kamchatka: Wilderness at the Edge*. China: Toppan Printing Co., 2008.

Steller, Georg. *Steller's History of Kamchatka: Collected Information Concerning the History of Kamchatka, Its Peoples, Their Manners, Names, Lifestyles, and Various Customary Practices*. Fairbanks, AK: University of Alaska Press, 2003.

Stockner, John. *Nutrients in Salmonid Ecosystems: Sustaining Production and Biodiversity*. Bethesda, MD: American Fisheries Society, 2003.

Taylor III, Joseph E. *Making Salmon: An Environmental History of the Northwest Fisheries Crisis*. Seattle, WA: University of Washington, 1999.

Walton, Izaak. *The Compleat Angler*. Norwalk, CT: The Heritage Press, 1976.

Whitworth, J. S.; Daniel, J.; Rubin, R.; Robbins, W. G; Woody, E.; Duncan, A.; Kitzhaber, J.; and twenty-one more. *Oregon Salmon: Essays on the State of the Fish at the Turn of the Millennium*. Portland, OR: The Irwin-Hodson Company, 2001.

Wilson, Edward O. *The Diversity of Life*. Cambridge, MA: Harvard University Press, 1992.

Wolfensohn, James D. *A Global Life: My Journey Among Rich and Poor, from Sydney to Wall Street to the World Bank*. New York, NY: Public Affairs, 2010.

ABOUT THE AUTHOR

TUCKER MALARKEY is the author of the critically acclaimed and national bestselling novels *An Obvious Enchantment* and *Resurrection*. Her career began at *The Washington Post,* and she was co-writer and researcher for *Sleepwalking Through History,* a bestselling account of the Reagan years by Haynes Johnson. Her love of wilderness and travel have taken her all over the world; she lived in Africa for years before attending the Iowa Writers' Workshop for fiction. She has since taught English and writing and worked as an editor and consultant for various literary magazines, writers, and filmmakers. She lives with her son in Berkeley, California, where she hosts visiting international filmmakers, writers, and artists in a creative collective that fosters mentoring as well as collaboration. Active in International Aid work and conservation, Tucker is currently working with a team of Stanford and Harvard scientists on a story about science, intuition, and the power of human connection as it relates to global health and conservation on the island of Madagascar. Malarkey grew up fly-fishing, studied Sovietology, and has traveled to Russia numerous times. She still fishes on the Deschutes River with her cousin Guido Rahr. *Stronghold* is her first major work of nonfiction.

ABOUT THE TYPE

This book was set in Sabon, a typeface designed by the well-known German typographer Jan Tschichold (1902–74). Sabon's design is based upon the original letterforms of sixteenth-century French type designer Claude Garamond and was created specifically to be used for three sources: foundry type for hand composition, Linotype, and Monotype. Tschichold named his typeface for the famous Frankfurt typefounder Jacques Sabon (c. 1520–80).